International Conference on Manufacturing Technologies (ICMT 2020)

IOP Conference Series: Materials Science and Engineering
Volume 842

Seattle, Washington, USA
17 - 19 January 2020

ISBN: 978-1-7138-2806-8
ISSN: 1757-8981

Printed from e-media with permission by:

Curran Associates, Inc.
57 Morehouse Lane
Red Hook, NY 12571

Some format issues inherent in the e-media version may also appear in this print version.

Institute of Physics
Dirac House, Temple Back
Bristol BS1 6BE UK

Phone: 44 1 17 929 7481
Fax: 44 1 17 920 0979

techtracking@iop.org

Additional copies of this publication are available from:

Curran Associates, Inc.
57 Morehouse Lane
Red Hook, NY 12571 USA
Phone: 845-758-0400
Fax: 845-758-2633
Email: curran@proceedings.com
Web: www.proceedings.com

TABLE OF CONTENTS

COMPOSITE MATERIALS

MATERIAL CHARACTERIZATION

PETROCHEMISTRY

MATERIAL PROCESSING

MECHANICAL ENGINEERING

IOP Conf. Series: Materials Science and Engineering **842** (2020) 011001 doi:10.1088/1757-899X/842/1/011001

Preface

The 2020 4thInternational Conference on Manufacturing Technologies (ICMT 2020) and 2020 8th International Conference on Nano and Materials Science (ICNMS 2020) were held in Seattle, USA on January 17-20, 2020. The conference program consisted of invited, oral, and poster presentations from scientists working in the areas of materials science and manufacturing technologies. In particular, the conference covered recent trends and advances in the field of Manufacturing Technologies.

The major goal of the conference was to bring academic scientists, engineers and industry researchers together to share and exchange their experience and research results, and to explore opportunities for collaboration. Distinguished Professors and experts delivered the latest state of the art information in their respective fields of expertise as well discussed the challenges and future directions.

The conference focused on important research issues associated with science and technology in the rapidly evolving fields of Manufacturing Technologies, Nano and Materials Science and to promote synergy between basic research and technological needs for real industrial applications. This event was intended not only as a vehicle for the dissemination of research results on the latest advances in the area of Manufacturing Technologies, Nano and Materials Science, but also as an opportunity for scientists from around the world to discuss developments in their respective countries, exchange ideas and explore the possibility for collaboration at international level. The wide ranging presentations at the conference included invited talks, oral and poster presentations, delivered over six sessions.

The conference proceedings consist of 26 selected papers presented at the conference. We hope that the conference proceedings will offer the readers a good insight into the future direction of research in this field of materials science and engineering.

We wish to express our gratitude to members of the conference committee and reviewers who gave their valuable time and advice, which helped in selection of high quality papers from all over the world for the conference and for publication in the proceedings. We also thank all the authors and participants for contributing to the success of the conference and we truly hope to see them again at the future ICNMS & ICMT conferences.

Conference Chair

Prof. Ramesh K. Agarwal, Washington University in St. Louis, USA

On behalf of the Conference committee

Seattle, WA, USA

IOP Conf. Series: Materials Science and Engineering **842** (2020) 011002 doi:10.1088/1757-899X/842/1/011002

Conference Chairs

Prof. Ramesh K. Agarwal, Washington University in St. Louis, USA
Prof. Steven Y. Liang, Georgia Institute of Technology, USA

Program Committee Chairs

Prof. Jing Wang, University of South Florida, USA
Prof. Devki N. Talwar, Indiana University of Pennsylvania, USA
Prof. Xu Chen, University of Washington, USA
Prof. Yong Suk Yang, Pusan National University, Korea
Prof. Farhang Pourboghrat, The Ohio State University, United States

Technical Committees

Prof. Anselmo Alves Bandeira, Federal University of Bahia, Brazil
Prof. Mohd Rafie Bin Johan, University of Malaya, Malaysia
Prof. Fei Zhou, Nanjing University of Aeronautics and Astronautics, China
Assoc. Prof. Wenke Gao, Lanzhou University of Technology, China
Prof. Abhijit Chanda, Jadavpur University, India
Prof. Himadri Chattopadhyay, Jadavpur University, India
Prof. Velamurali, Anna University, India
Prof. A. Elaya Perumal, Anna University, India
Prof. N. V. Raghavendra, National Institute of Engineering, Mysuru, India
Prof. Recai KUS, Selcuk University, Turkey
Assoc. Prof. Prasanna Shakti Jena, Vardhaman College of Engineering, India
Dr. Kamran Shavezipur, Southern Illinois University Edwardsville, USA
Dr. Marco Castellani, University of Birmingham, UK
Dr. Marta Menegoli, Naica SC, Italy
Dr. Aydin Berenjian, The University of Waikato, Ireland
Dr. D. Ramasamy, Universiti Malaysia Pahang, Malaysia
Dr. Mohsen Motahari-Nezhad, Shahid Beheshti University
Dr. Hatem Mrad, Université du Québec en Abitibi-Témiscamingue, Canada
Dr. Chen-Yuan Chung, National Central University, Taiwan
Dr. Aydin Berenjian, The University of Waikato, Ireland

First Name	Last Name	Institution	Email Address
Steven Y.	Liang	Georgia Institute of	steven.liang@me.gatech.edu
Jianxiao	Bian	Longdong University,	Jxbian@ldxy.edu.cn
Jianxiao	Bian	Longdong University,	Jxbian@ldxy.edu.cn
Chen-Yuan	Chung	National Central	cxc474@case.edu
Mahmood Fawzi	Ahmed	University of	mahmodmfa81@gmail.com
Wimalin	Laosiritaworn	Chiang Mai	wimalin@hotmail.com
Kreangkrai	Maneeintr	Chulalongkorn	Krengkrai.M@chula.ac.th
Fang	Cheng	Zhejiang University,	fcheng@zju.edu.cn
Wenke	Gao	Lanzhou university of	gaowk_best@163.com
Imen	Bouzarkouna	Cesi Rouen & INSA	ibouzarkouna@cesi.fr
Rishi	Pahuja	University of	rpahuja@uw.edu
Ramulu	Mamidala	University of	ramulum@uw.edu
Y V	Sinitsyna	Bauman Moscow State	J.Sinitcyna@gmail.com
Q C	Zhou	Tongji University,	576947181@qq.com
Sotirios	Grammatikos	Norwegian University	Sotirios.grammatikos@ntnu.no
Sotirios	Grammatikos	Norwegian University	Sotirios.grammatikos@ntnu.no
Sotirios	Grammatikos	Norwegian University	Sotirios.grammatikos@ntnu.no
Bilal Anjum	Ahmed	National Univeristy of	banjumahmed@gmail.com
Bryce Marie Pettit	Estell	Georgia Institute of	bryce123pe@gmail.com
Sotirios	Grammatikos	Norwegian University	Sotirios.grammatikos@ntnu.no
David	Chavez	Universidad	david_chavezc@uadec.edu.mx
Sunchul	Huh	Gyeongsang National	schuh@gnu.ac.kr
David	Chavez	Universidad	david_chavezc@uadec.edu.mx
Shoroog	Alraddadi	Umm Alqura	swraddadi@uqu.edu.sa
Sunchul	Huh	Gyeongsang National	schuh@gnu.ac.kr
Kong Kian	Hau	National University of	ceekkh@nus.edu.sg

IOP Conf. Series: Materials Science and Engineering **842** (2020) 011004 doi:10.1088/1757-899X/842/1/011004

Peer review statement

All papers published in this volume of *IOP Conference Series: Materials Science and Engineering* have been peer reviewed through processes administered by the proceedings Editors. Reviews were conducted by expert referees to the professional and scientific standards expected of a proceedings journal published by IOP Publishing.

IOP Conf. Series: Materials Science and Engineering **842** (2020) 012001 doi:10.1088/1757-899X/842/1/012001

Characterization of Surfaces Generated in Milling and Abrasive Water Jet of CFRP Using Wavelet Packet Transform

Rishi Pahuja* and M Ramulu

University of Washington, Seattle WA

*E-mail: rpahuja@uw.edu

Abstract. Polymer matrix composites find extensive application in aerospace and automotive industry. Although composites are molded to a near net shape, secondary machining such as hole-drilling and edge trimming is unavoidable. Conventional milling and Abrasive water jet are widely employed processes to cater the machining needs. Because of the nature of material-tool interaction, both the processes have their limitations resulting in machining induced damages such as matrix cracking, exit-delamination, fiber pull-outs, spring-back, interply delamination, striations, abrasive embedment, and kerf taper.

Surface roughness plays a pivotal role in determining the damage induced during the machining processes. Traditionally roughness parameters – average roughness (R_a) and ten-point average (R_z) are used to characterize the surfaces. Often, the use of cut-off wavelengths, governed by standards, result in poor distinguishability for machined surfaces of anisotropic materials such as polymer matrix composites (PMC). The insensitivity and poor distinction in certain cases leads to unsound judgment in characterizing the machining induced damage, and also in the selection of process parameters. This necessitates other methods and indicators that better identify the surface signature and correlate with the process parameters. In this study, a novel approach is proposed in which wavelet packet transform (WPT) is applied to the machined surface profiles. An indicator (ratio of wavelet packet energy and entropy) is defined to characterize the surface and better predict the surface quality as a function of machine tool, material and process variables. Regardless of the type of CFRP and machining process (AWJ and conventional milling), WPT indicator proved to be a better predictor of the machining process ($R^2 > 80\%$) when compared with R_z ($R^2 < 67\%$).

1. Introduction

Although composite materials (CFRP) are molded to a near net shape, secondary machining is often required to meet the dimensional tolerances and attain desired functional performance [1,2]. Two primary processes in practice are conventional trimming and abrasive water jet (AWJ) machining. The anisotropy of CFRP laminates decide the resistance offered to the machine tool, which in turn governs the surface quality. In conventional milling, because of the abrasive nature of fibers, tool wear is prominent. Besides, matrix cracking, exit-delamination, fiber pull-outs, spring-back and interply delamination limits the process speed and affects the process viability [3]. Abrasive Water jet is a high speed alternative to conventional milling, but striations due to jet energy loss, abrasive embedment, exit delamination and kerf taper are a few limitations [4–6].

The damages and defects introduced in CFRP during machining could be reflected on the surface or sub-surface. If unaccounted in the component design, these defects could severely affect the static and dynamic mechanical properties [7]. The relationship between the surface features reflecting these

IOP Conf. Series: Materials Science and Engineering **842** (2020) 012001 doi:10.1088/1757-899X/842/1/012001

defects and process parameters is important for a defect free machining. Traditionally, gaussian filter is used to calculate roughness parameters, however the technique is computationally expensive and often, more than one parameter is required for wholesome characterisation of the engineering surfaces.
Wavelet analysis is has proven to be an effective tool in signal analysis. Lee et. al. [8] were among the first to propose the morphological characterisation of engineering surfaces using Wavelet transform. Kim et. al studied correlated the surface quality of steel with wavelet energy signatures [9]. Wang et. al. [10] used Wavelet Packet transform for feature extraction and roughness evaluation. There is a dearth of literature in studying the surfaces machined with Abrasive Water Jet (AWJ), especially for challenging materials such as prepreg-based CFRP. Pahuja et al. [11] used discrete wavelet analysis to study the surfaces of AWJ machined carbon foam.
In this study, a novel approach is proposed where wavelet packet transform (WPT) alongwith a surface feature indicator – energy-entropy coefficient is used to characterize the machined surfaces of prepreg-based, random chopped discontinuous fiber composite – HexMC. The random and inhomogeneous nature of the material results in localised damage, necessitating effective techniques to identify these damages [12,13]. Upon comparison with traditional roughness parameters, WPT energy-entropy signature was found to be superior in describing the surface texture, and its correlation with the machining process used to generate those surfaces.

2. Materials, Methods and Equipment

In this study, random chopped discontinuous fiber composite (DFC), commonly known as HexMC® was used. HexMC is made up of prepreg chips of size 50.8 mm x 8 mm chopped, randomly distributed and consolidated through molding process. The nominal thickness of HexMC laminate was 11 mm, and mechanical properties are given in Table 1. The SEM micrograph in **Figure 1** shows the random nature of prepreg chips in the laminate.

Table 1. Mechanical properties of DFC – HexMC [12]

Property	
Tensile Strength	300 MPa
Tensile Modulus	38 GPa
Flexural Strength	500 MPa
Flexural Modulus	30 GPa
Compression strength	290 MPa
Compression Modulus	38 GPa
Interlaminar shear strength	250 MPa
Interlaminar shear modulus	15 GPa
Poisson's ratio	0.3

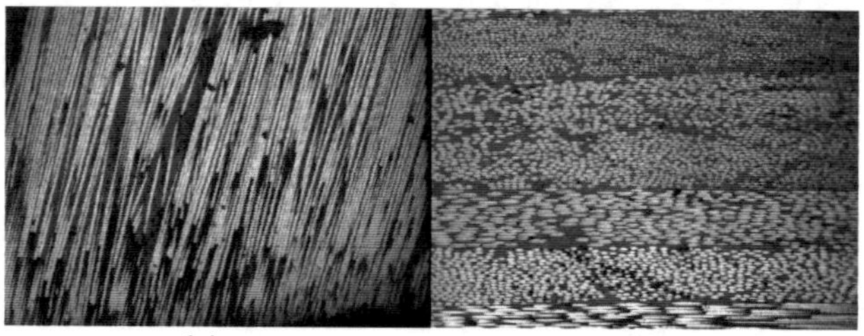
(a) (b)

Figure 1. SEM micrograph of HexMC (a) Top view, (b) Side view.

The conventional milling experiments were performed using a 3 axis Haas milling machine DFC laminate was edge trimmed using a two-flute carbide tool of diameter 6.35 mm and helix angle 30°. Tool was advanced 1 mm radially into the material and through-the-thickness in axial direction. The length of cut (straight) was 50 mm. The Abrasive Water Jet experiments were performed using Flow International® water jet machine equipped with 400 MPa intensifier. Details about the experimental conditions can be found in the previous work [12]. The process variables and their levels for conventional milling and AWJ are given in Table 2.

Table 2. Experimental variables and their levels for conventional and AWJ machining.

Conventional milling		**Abrasive Water Jet**	
Spindle speed	1000 and 6000 rpm	Pressure	250, 275, 350 MPa
Feed rate	5, 10, 15, 20, 25 mm/	Abrasive flow rate	5.3, 6, 6.8 g/s
		Traverse speed	5, 10, 15, 19 mm/s

IOP Conf. Series: Materials Science and Engineering **842** (2020) 012001 doi:10.1088/1757-899X/842/1/012001

The surface roughness profile was measured using contact type roughness measuring equipment (Mahr XR20). With a skidless probe of 2 μm diameter. The profile evaluation length was 5.6 mm with sampling length of 0.8 mm. Gaussian filtering with cut-off length 2.5 μm was used to calculate the ten point average R_z roughness parameter. Several roughness profiles were measured for each machined surfaces, with a distance of h=1,2,…k mm (8 and 11 for conventional and AWJ respectively) from the top edge. Multiple measurements were made to increase the probability of capturing severe localized damage, if present. Overall, 376 profiles were evaluated in this study.

3. Analysis methodology

Wavelet packet tranform (WPT) decomposes the signal onto time and frequency scales. Starting with a mother wavelet, the signal is decomposed such that it is downsampled by 2, resulting in two subbands– approximation and detailed coefficents. The successive decompsosition(s) is carried out by scaling and translation of the wavelet function, on both approximation and detailed subbands. If the original sampling frequnecy of the signal is f_s, the first level decomposition will result in 2 subbands with

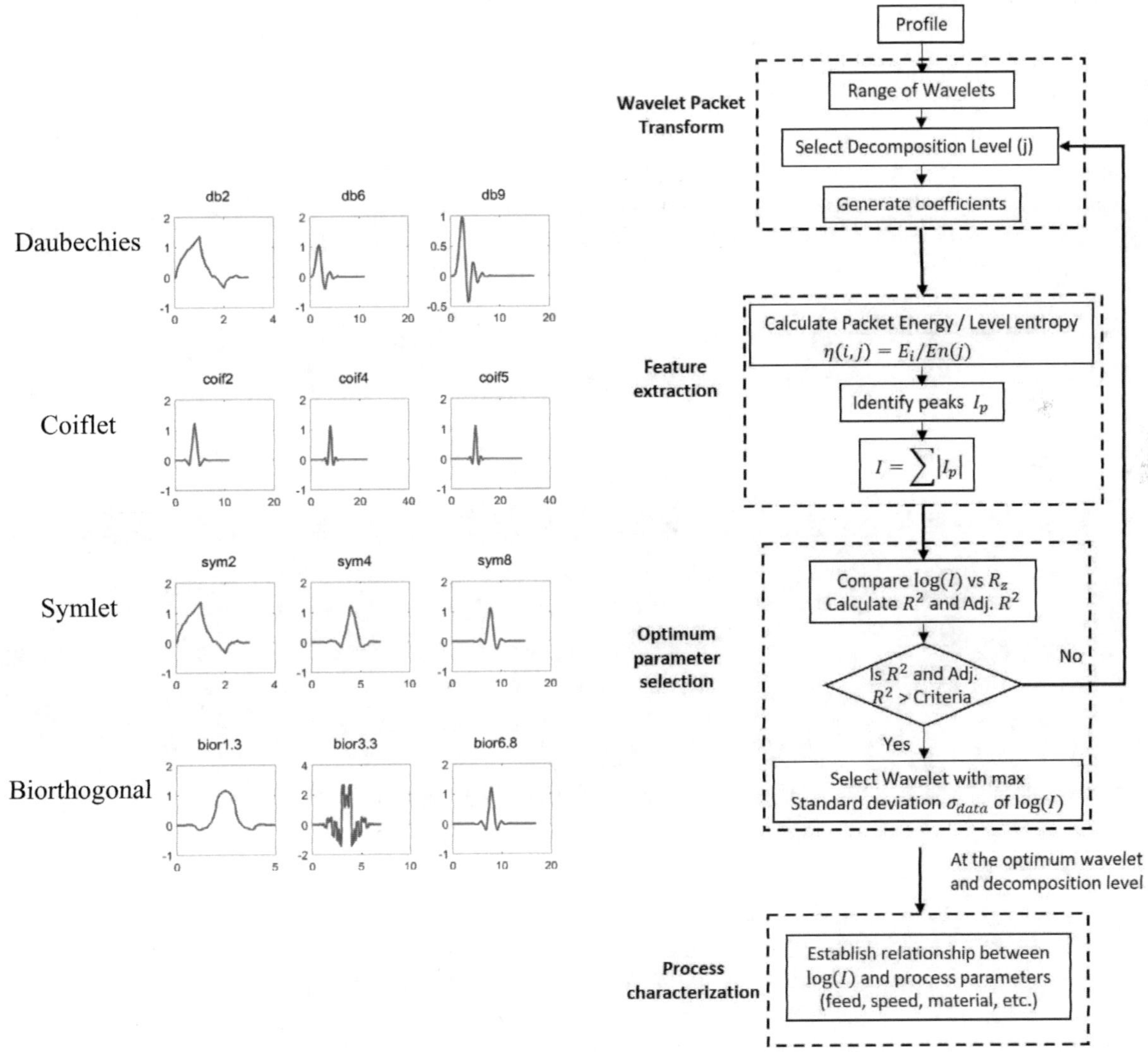

(a) (b)

Figure 2. (a) Shape of mother wavelets, (b) Wavelet Packet Transform (WPT) analysis algorithm.

frequency ranges [0, 0.25f_s) and (0.25f_s, 0.5f_s]. At j^{th} decomposition level, 2^j subbands (or wavelet packets) are generated.

The surface profiles were evaluated using Wavelet packet transform (WPT). Usually, mother wavelets are used based on their phenomonological significance as reflected in the signal. A wavelet which best describes the nature of the signal, will extract the signal features in the most effective manner. In this study, 35 mother wavelets have been selected for the analysis, as in [14] and depicted in Figure 2(a). The first step is the wavelet packet transform of the roughness profile. For a given decomposition level, the wavelet packets were calculated and arranged in ascending frequency order. Energy of each wavelet packet at the given decomposition level is given computed. The total energy at each decomposition level is given by the summation of energy of all packets at that level. Next, Shannon entropy was calculated. Further, the ratio of packet energy and total Shannon entropy was determined as given by equation (1).

$$\eta(i,j) = \frac{E_i}{En_T} \tag{1}$$

This energy-entropy coefficient (η) was then plotted against the frequency ordered wavelet number, and dominant peaks were identified. The summation of the absolute values of these peaks identified as a unique classifier for a given profile. This WPT-indicator, *I*, is given by equation (2).

$$I = \sum \eta_{peaks} \tag{2}$$

The optimal class and order of mother wavelet was identified as given in Figure 2(b). Upon selection of optimum wavelet and decomposition level, the profiles were correlated with the process variables. The wavelet packet analysis algorithm is summarized in Figure 2(b).

4. Results and Discussion

4.1. Conventional milling

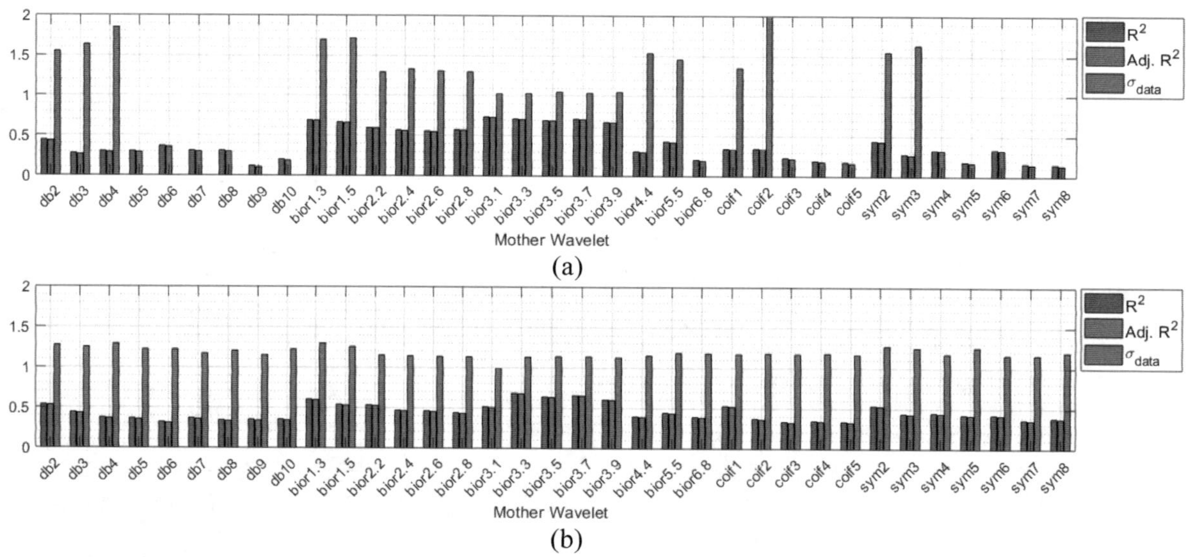

Figure 3. R^2, Adjusted R^2 and Standard deviation (σ) for 35 wavelets at decomposition level (a) Level 5 and (b) Level 10 for conventionally milled HexMC.

Figure 3 shows R^2, adjusted R^2 and standard deviation for regression between WPT Indicator *I* and R_z. The Biorthogonal wavelets outperformed other wavelets for nearly all decomposition levels. Bior3.3 at decomposition level 7 was selected with R^2, adjusted R^2 and standard deviation of 72.25%, 71.78% and 0.9169 respectively. Although R^2 is low, it was the best that was achieved using any wavelet and decomposition level. As shown in Figure 3(a), the performance at level 5 was poor and many cases were eliminated in calculation due to non-existence of the Indicator *I*. At the selected level (Level-7), the wavelets – db2, coif1 and sym2 were better among their respective families of wavelet. However, R^2 was less than 40% for these mother wavelets. The maximum value of log *I* for Bior3.3 at level 7 is 5.4128 which is 18.9%, 20.6% and 18.9% higher than db2, coif1 and sym2 respectively. Figure 4 shows a strong positive correlation between R_z and *I* for Bior3.3 wavelet among others.

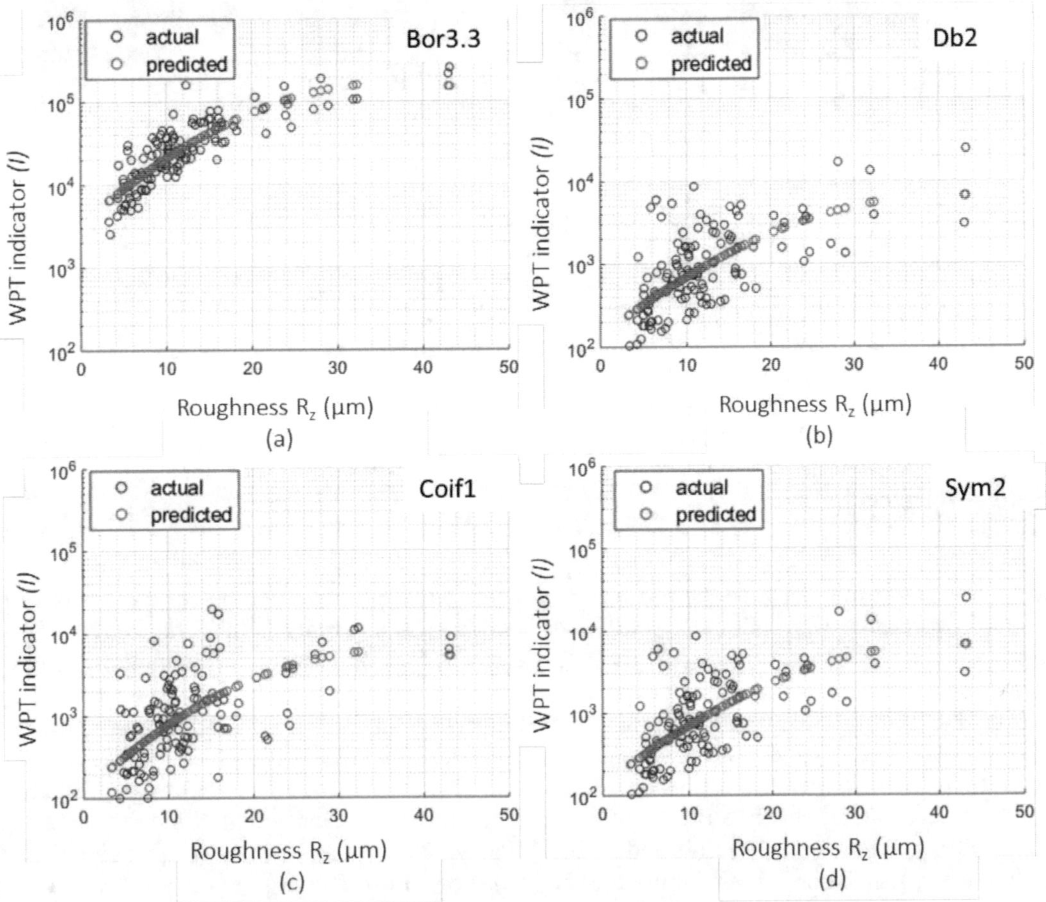

Figure 4. Relationship between *I* and R_z on a log-linear scale for conventionally milled HexMC profiles analyzed using mother wavelet – (a) Bior3.3, (b) Db2, (c) Coifl and (d) Sym2 at level 7 decomposition.

Figure 5 shows the correlation between WPT indicator *I* and process parameters. In general, a positive trend between feed and *I* was observed. At high spindle speed (6000 rpm), the magnitude of indicator (*I*) increased from 3×10^4 to 9.3×10^4 for specimens machined with feed ranging from 5-20 mm/s. A sudden reduction in *I* was observed at 25 mm/s feed. At 1000 rpm spindle speed, the resistance to cutting was higher and resulted in high *I* value. The magnitude of *I* ranged between 8×10^4 and 25.9×10^4.

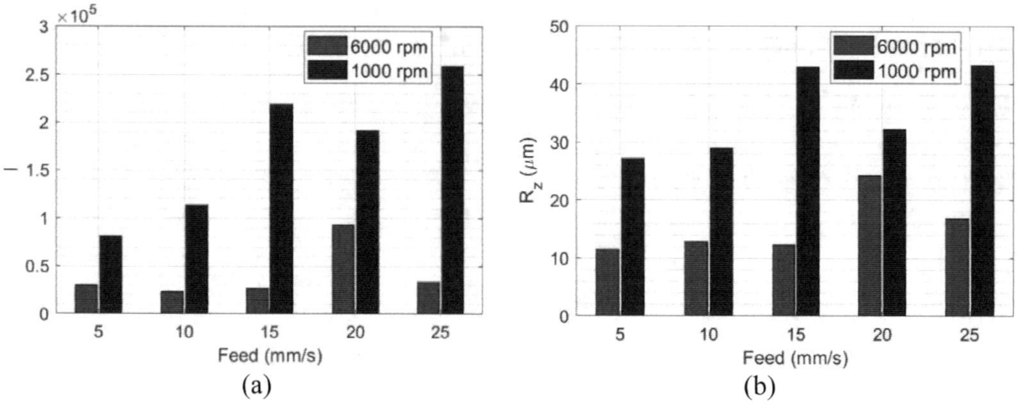

Figure 5. Effect of feed and spindle speed on surface quality expressed in terms of WPT indicator (*I*).

4.2. *Abrasive Water Jet machining*

The abrasive process involves jet-material interaction in such a way that the energy of the jet is dissipated as the jet penetrates through the workpiece. If the power of the jet (i.e. abrasive velocity) is low or the exposure time of the jet is small, the jet may deflect back due to the loss of energy and hence its cutting ability. The beginning of the curving of jet cutting front is usually specified as the beginning of rough cutting region (RCR) and often needs to be minimized to meet the tight surface roughness and dimensional tolerances.

The wavelet analysis, similar to conventional milled surfaces, resulted in optimal wavelet– Bior3.3 and decomposition level 8. Next, the relationship between process parameters and WPT indicator *I* was studied. The relationship between the process parameters water pressure (P), speed (u), Abrasive Flow rate (AFR) and response variables – R_z and *I* was developed using Analysis of Variance (ANOVA). A cubic polynomial model was implemented with F-value of 36.24. Backward elimination procedure was used to eliminate the insignificant factors by comparing the f-values with the critical f-ratio at alpha = 0.05 level. Pressure and feed were the most contributing parameters in the model. A high pressure resulted in high jet power and hence a smoother surface was obtained. A high feed rate resulted in low exposure time, low jet energy per unit area and hence rougher surface. Figure 6 shows the effect of pressure and feed rate at abrasive flow rate 6 g/s. When measured near jet entry side (h= 1 mm), the maximum R_z was 28 μm at u=19 mm/s and P=200 MPa, which was 62.3 μm near jet exit side (h=8 mm) and at the same process conditions. Figure 6(c) and (d) shows the effect of pressure and feed rate on Rz at h=1 mm and h=8 mm respectively. The maximum value of *I* was 3.3 x 10^5 at jet entry side (h=1 mm) at u=19 mm/s and P=19 mm/s as shown in Figure 6(a). The WPT indicator *I* was increased to 2.36 x 10^6 when measured at h=8mm which is an indicator of striated geometry. In comparison, the increment in R_z and *I* from h=1mm to h=8mm was 122.1% and 617.1% respectively. This indicates that the poor surface quality is more accentuated by *I* as opposed to R_z, making I to be a better indicator to represent surface quality.

IOP Conf. Series: Materials Science and Engineering **842** (2020) 012001 doi:10.1088/1757-899X/842/1/012001

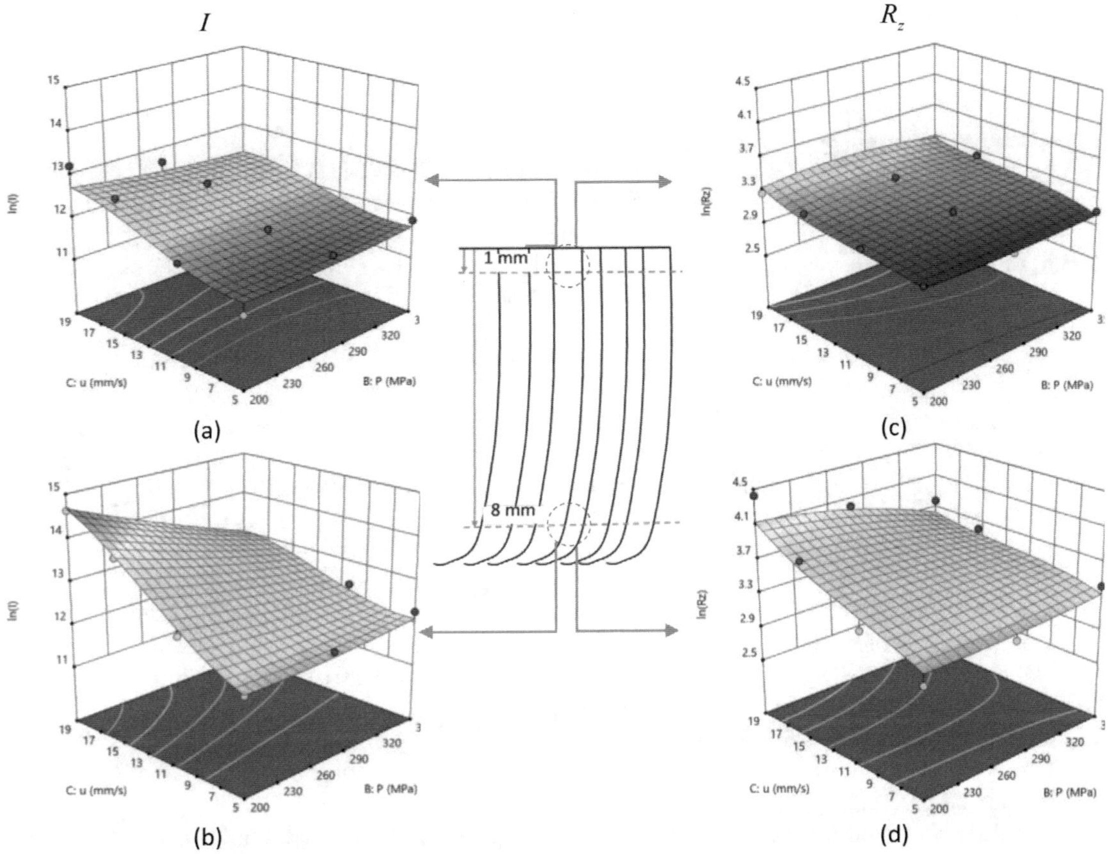

Figure 6. Effect of pressure (P - MPa) and jet feed rate (u-mm/s) on R_z at jet penetration distance (a) h=1 mm, (b) h=8 mm, and on I at jet penetration distance (a) h=1 mm, (b) h=8 mm.

5. Summary and Conclusion

In this study, Wavelet Packet Transform (WPT) was used to identify characteristic features of surface profiles generated by different machining processes and parameters. HexMC, a random chopped discontinuous fiber composite (DFC) was machined using surface conventional milling and Abrasive Water jet machining process. Wavelet coefficients were calculated after decomposing the surface profiles at different scales. A novel approach was proposed to characterize the process using the ratio of energy and entropy of dominant wavelet packets (*I*). The selection of mother wavelet and decomposition level was carried out by comparing the WPT indicator '*I*' with ten-point roughness parameter (R_z). For both AWJ and conventional milling, WPT indicator *I* proved to be a better predictor of process conditions when compared to R_z. Irrespective of the type of composite material and process used to generate the surface profile, higher value of *I* indicated poor surface quality. Mother wavelet Bior3.3 was found to be optimal for both machining processes. Decomposition levels 7 and 8 resulted in a reasonable correlation for conventional and AWJ milling respectively. A low spindle speed and high feed resulted in high R_z and *I* values. The surfaces generated by Abrasive Water jet showed higher value of R_z and *I*. Low pressure and high speed resulted in high R_z and *I*. Predictive models were developed to correlate *I* and process parameters. A comparison between WPT indicator *I* revealed high correlation of *I* with the process parameters (R^2>80%) as opposed to R_z (R^2<67%)

IOP Conf. Series: Materials Science and Engineering **842** (2020) 012001 doi:10.1088/1757-899X/842/1/012001

WPT method alongwith the proposed indicator– ratio of packet energy to Shannon was developed as an ideal procedure to define and characterize the surface quality; identify machining induced damage in composites; and correlate the surface integrity with process conditions.

Acknowledgments
Authors wish to acknowledge the encouragement and advice of Dr. M. Hashish. This work was supported by Boeing-Pennell professorship funds.

References
[1] Abrate S and Walton D A 1992 Machining of composite materials. Part I: Traditional methods *Compos. Manuf.* 3 75–83
[2] Abrate, S.; Walton D 1992 Machining of composite materials. Part II : Non-traditional methods *Compos. Manuf.* 3 85–94
[3] Ramulu M 1997 Machining and surface integrity of fibre-reinforced plastic composites *Sadhana* 22 449–72
[4] Hashish M 1993 Precision drilling of composites with abrasive-waterjets *Am. Soc. Mech. Eng. Mater. Div. MD* 45 217–25
[5] Hashish M 1989 Machining of advanced composites with abrasive-waterjets *Manuf. Rev.* 2 142–50
[6] Alberdi A, Suárez A, Artaza T, Escobar-Palafox G A and Ridgway K 2013 Composite cutting with abrasive water jet *Procedia Engineering* vol 63 pp 421–9
[7] Haddad M, Zitoune R, Bougherara H, Eyma F and Castanié B 2014 Study of trimming damages of CFRP structures in function of the machining processes and their impact on the mechanical behavior *Compos. Part B Eng.*
[8] Lee S H, Zahouani H, Caterini R and Mathia T G 1998 Morphological characterisation of engineered surfaces by wavelet transform *Int. J. Mach. Tools Manuf.*
[9] Kim D, Han C and Liu J J 2009 Optimal wavelet packets for characterizing surface quality *Ind. Eng. Chem. Res.* 48 2590–7
[10] Wang X, Shi T, Liao G, Zhang Y, Hong Y and Chen K 2017 Using Wavelet Packet Transform for Surface Roughness Evaluation and Texture Extraction *Sensors* 17 933
[11] Pahuja R and Ramulu M 2019 Study of surface topography in Abrasive Water Jet machining of carbon foam and morphological characterization using Discrete Wavelet Transform *J. Mater. Process. Technol.*
[12] Pahuja R and Ramulu M 2016 Machinability of Randomly Chopped Discontinuous Fiber Composites: A Comparative Assessment of Conventional and Abrasive Waterjet *23rd International Conference on Water Jetting 2016, Seattle, USA* (Seattle: BHR Group) pp 127–48
[13] Feraboli P, Peitso E, Deleo F, Cleveland T and Stickler P B 2009 Characterization of prepreg-based discontinuous carbon fiber/epoxy systems *J. Reinf. Plast. Compos.* 28 1191–214
[14] Pahuja R and Ramulu M 2019 Surface quality monitoring in abrasive water jet machining of Ti6Al4V–CFRP stacks through wavelet packet analysis of acoustic emission signals *Int. J. Adv. Manuf. Technol.* 104 4091–104

IOP Conf. Series: Materials Science and Engineering **842** (2020) 012002 doi:10.1088/1757-899X/842/1/012002

On the Mechanical Recycling of Decommisioned Insulation Polymer Composite Components

Sotirios Grammatikos[1,a)], Spyros Tsampas[2,b)], Styliani Papatzani[3,4,c)], Tang Luping[5,d)], Ingemar Löfgren[6,e)], Jocke Petterson[7,f)]

[1]Group of Sustainable Composites, Department of Manufacturing and Civil Engineering, Norwegian University of Science and Technology, Gjøvik, Norway
[2]R&T Centre, GKN Aerospace Engine Systems, Sweden
[3]Hellenic Army Academy, Department of Mathematics and Engineering Sciences, Vari, Greece
[4]Hellenic Ministry of Culture, Directorate of Restoration of Medieval & Post-medieval Monuments, Athens, Greece
[5]Department of Architecture and Civil Engineering, Chalmers University of Technology, Sweden
[6]Thomas Concrete Group AB, Sweden
[7]Material Technologies, Swerea SICOMP AB, Sweden

[a] sotirios.grammatikos@ntnu.no

Abstract. Fibre reinforced polymer composites (FRPs) are being increasingly used in aerospace and automotive applications due to their high specific mechanical properties. The construction industry has also started taking advantage of the potential of FRPs for both structural and non-structural purposes. The result of this remarkable absorption of FRPs within the worldwide production market, has led to an immense increase of decommissioned thermoset-matrix components. Nowadays, the majority of the decommissioned FRP components are recovered energy-wise through incineration or simply discarded in landfills around the globe. Within the framework of this paper, we present a solution for the extension of the service life of decommissioned FRP components. Decommissioned electrical insulation FRP pipes were granulated and incorporated as fillers within both cementitious and polymer matrix composites. The effect of FRP granulates on the mechanical performance of cementitious and polymer matrix composites is examined to determine the maximum granulate-filler fraction that can be recycled without compromising the mechanical performance and manufacturing process.

1. Introduction

The majority of fibre-reinforced composites used in industries such as aeronautics, energy and transportation once they reach their out-of-service life, end up in landfill or are being used for energy production mainly via incineration (according to The Incineration of Waste Directive, 2000/76/EC) [1]. In particular, every year in Europe more than 60 000 tonnes of composite waste is produced of which most ends up in landfill around Europe or other non-EU countries (EC Waste shipment regulations (259/93/EEC)) [2].
Although incineration and disposal to landfill are very detrimental for the environment, the low cost of glass fibres and lack of commercially viable industrial recycling methods make these solutions

inevitable. Recently, the use of commercially manufactured fibre-reinforced composites in concrete materials has been studied by a number of researchers as a reinforcement phase in concrete [3-6]. In particular, these studies have considered fibre-reinforced composites ground into powder or reinforcing bars (rebar) cut into aggregate-sized pieces as filler/reinforcement for concrete with promising results in terms of processability and mechanical properties.

On the contrary, this work presents an investigation towards a more sustainable recycling solution for fibre-reinforced composites particularly those used in the energy industry. That said, a feasibility study to assess the potential of recycling/reuse of glass fibre-reinforced polymer (GFRP) composite insulators and discarded glass reinforcements was carried out. The concept-idea was to incorporate decommissioned epoxy-based glass fibre-reinforced insulators as fillers in i) cement-based products for potential replacement of gravel/crushed stones and ii) polymer-based products for potential replacement of reinforcing fillers i.e. glass fibre reinforcement. After a thorough investigation which involved workability, manufacturing and mechanical testing, it was concluded that discarded FRPs can be efficiently recycled (downcycling) within cementitious and polymer matrix products at 3%v/v and 25%w.w, respectively, without compromising their original structural performance.

2. Experimental

2.1. Materials

The decommissioned electrical insulating FRP tubes used in this study were supplied by ABB Composites AB, Sweden. The configuration of the tubes is illustrated in figure 1. The studied FRP tubes consisted of epoxy resin and E-glass fibres. The tubes were fabricated using filament-winding allowing for a 70% fibre volume fraction. After shredding, FRP granules where incorporated as fillers in cementitious and polymer matrix composites. The FRP-filler modified cementitious and polymer composites were subsequently structurally evaluated using standard mechanical testing (compression, 3-point bending) to assess their performance compared to the original material (no FRP-filler addition).

Figure 1. Decommissioned FRP insulation tubes from ABB Composites AB that were ground and used as filler in cementitious and polymer matrix composites.

2.2. *Testing methodology*

2.2.1. Mechanical recycling – shredding. The FRP tubes were converted into fillers through shredding employing a granulator provided by RAPID Granulator, Sweden, AB. The shredded fillers were filtered through a 6 mm metallic mesh.
2.2.2. Mechanical testing. Cementitious composites (mortars) were tested under compression and 3-point bending tests to evaluate the compressive and flexural strength, respectively. Compression and flexure tests were performed according to the BS EN 1015-11:1999 standard for testing. Both reference and FRP-filler modified specimens were tested with three replicates in all cases.

3. Results and discussion

3.1. *Cementitious composites (mortars)*
Standard cementitious mortars (1350 g sand, 450 g cement, and 225 g water) were fabricated to study the effect of FRP fillers in mortars.

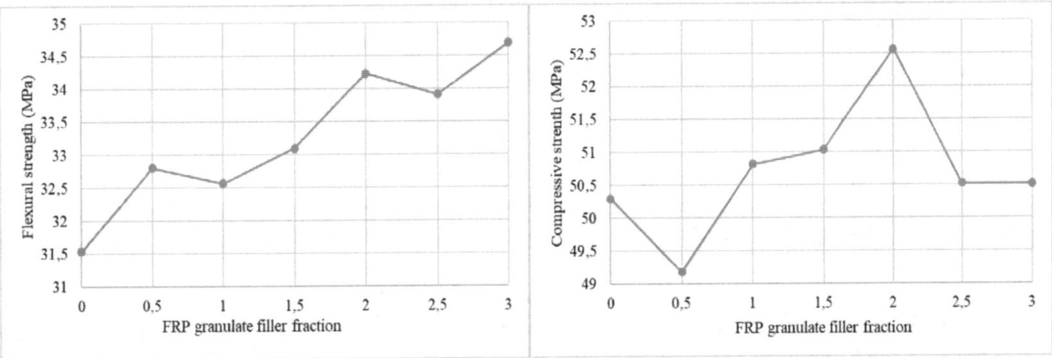

Figure 2. Flexural strength (left) and compression strength (right) for the different FRP granulate fractions.

In order to reach the goal of the study, FRP fillers were added to a certain degree or until workability of the cementitious mix was affected. To that respect, a maximum amount of 3% by volume of FRP ground fillers was added within the standard mortar mix. A gradual increase of the FRP fillers content was adopted to effectively test filler influence on the cementitious mix. Finally, prismatic mortar specimens (40x40x160mm) containing 0%, 0.5%, 1%, 1.5%, 2%, 2.5% and 3% of FRP filler were fabricated. Figure 2 depicts the mean values of flexural and compression strength for the different FRP filler fractions. As can be seen, flexural strength increases marginally with an increase in FRP fraction whereas compressive strength remains practically unaffected. However, 3%v/v was the fraction point where workability of mortars was significantly influenced.

3.2. *Polymer composites*
Composite laminates consisting of a mixture of ground FRP and new raw material (polymer and fibres) were produced and tested in 3-point bending testing. The laminates consisted of vinylester polymer for the matrix and E-glass glass fibres for the reinforcement together with 25%w/w recycled FRP fillers. Two types of laminates were fabricated (namely S1 and D1) using 167.89 gr (S1) and 260.59 gr (D1) of glass reinforcement, respectively. Standard wet hand-lay-up at room temperature conditions was employed for the manufacturing. Three-point bending tests were conducted in accordance with the ASTM D7264 having a support span of 150 mm. The dimensions of the test samples were 180x25 mm. Figure 3 displays the 3-point bending testing setup whereas figure 4 representative load deflection curves for the two fabricated types of specimens.

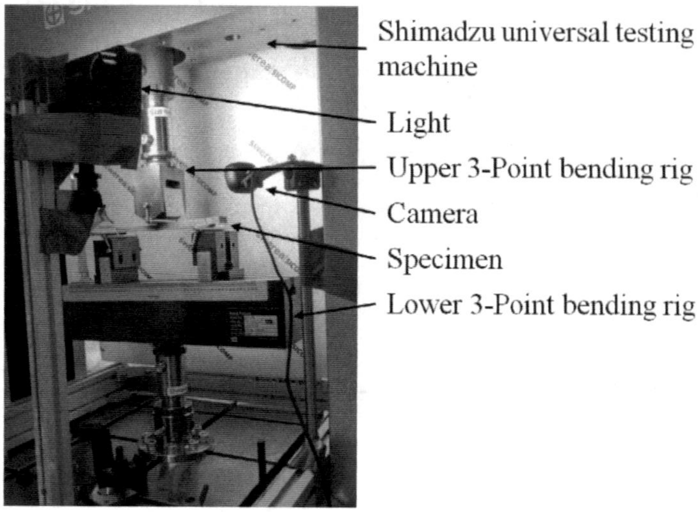

Figure 3. Three-point bending setup.

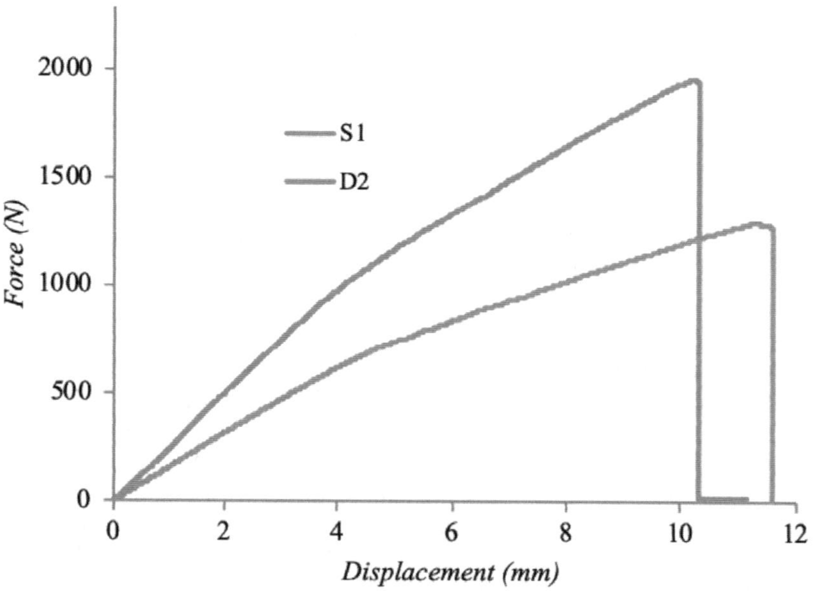

Figure 4. Load/Deflection curve from specimen S1 and D2.

4. Conclusions

In this study, it was demonstrated that FRP composite insulators can be ground down and form functional/valuable additives for both cementitious and polymer composite products. Commercial drive for introducing this type of recycling is currently considered relatively low, as the material fractions replaced with recycled material are relatively low, thus reducing the economic impetus for introduction. However, considering the environmental effects when landfilling such FRP components and the release of toxic substances when decomposing, this work significantly contributes towards the development of measures for recycling thermosetting polymer composites. From a structural point of view, FRP granulates can be employed as efficient reinforcing fillers in cementitious and polymer composites up to 3%v/v and 25% w/w, respectively, without compromising their original mechanical behaviour. In the case of polymer composites, mechanical performance is somewhat lower than for most laminates used commercially today but high enough for many non-structural applications. The results demonstrate simple process technology for cementitious mortars and polymer composite laminates that can be applied to a variety of construction applications such as pavements, bridge decks, wall panels, boats, etc. It becomes obvious that this feasibility study proved the potential of recycling/repurposing of retarded FRP insulation pipes contributing to the waste reduction, minimization of CO_2 emissions, efficient use of resources, etc.

5. References

[1] European Commission, Directive 2000/76/EC of the European Parliament and of the Council on the incineration of waste, 2008.

[2] European Commission, Report from the Commission to the European Parliament and the Council on the implementation of regulation (EC) no 1013/2006 on shipments of waste, 2015.

[3] N. Reynolds, M. Pharaoh, An introduction to composites recycling, in: V. Goodship (Ed.), Management Recycling and Reuse of Waste Composites, Woodhead Publishing, 2010, pp. 3–19.

[4] M. Shahria Alam, E. Slater, Billah.A. Muntasir, Green concrete made with RCA and FRP scrap aggregate: fresh and hardened properties, J. Mater. Civ. Eng. 25 (12) (2013) 1783–1794.

[5] M.C.S. Ribeiro, A.C. Meira-Castro, F.G. Silva, J. Santos, J.P. Meixedo, A. Fiúza, Reuse assessment of thermoset composite wastes as aggregate and filler replacement for concrete-polymer composite materials: a case study regarding GFRP pultrusion wastes, Resour. Conserv. Recycl. (2015).

[6] A. Yazdanbakhsh LC. Bank, C. Chen, Use of recycled FRP reinforcing bar in concrete as coarse aggregate and its impact on the mechanical properties of concrete, Construction and Building Materials 121 (2016) 278–284.

Acknowledgments

This work was funded by RECYTAL project (Vinnova/Energimyndigheten/Formas (Sweden) - RESource utlysning 2 (2016-2017), Project reference: 2017-001986). Authors are thankful to Marek Machowski (Chalmers University of Technology) and Peter Hellström (Swerea Sicomp) for their contribution to experimental testing. ABB Composites, Sweden, is greatly acknowledged for supplying discarded insulation FRP tubes whereas RAPID Granulator, Sweden, for shredding FRP tubes into fragments.

IOP Conf. Series: Materials Science and Engineering **842** (2020) 012003 doi:10.1088/1757-899X/842/1/012003

Effect of The Dissolution Method With N, N-Dimethylformamide As A Solvent On The Electrical Properties Of Acrylonitrile Butadiene Styrene And Carbon Nanotube Composites

Bryce M. Pettit Estell [1, a] **and Rosario A. Gerhardt** [1,b]

[1]School of Materials Science and Engineering, Georgia Institute of Technology, 771 Ferst Drive Atlanta, GA 30332-0245

[a] bryce123pe@gmail.com [b] Rosario.gerhardt@mse.gatech.edu

Abstract. The objective of this study is to create electrically conducting composite polymers using the dissolution method. Composites are made from Acrylonitrile Butadiene Styrene (ABS) as the matrix and Carbon Nanotubes (CNTs) as the filler. Composites were made using the dissolution method with N, N-Dimethylformamide (DMF). All the components were mixed then dried in a fume hood until they were formed into pellets in a compression molder. Compressed samples were optically and electrically tested. Results show that composites that were only dried in the fume hood have a bubbled structure and reach percolation around 1-5 phr. The bubbles were determined to be coming from water that got into the samples while they were drying in the fume hood. Composites were placed in a furnace at 100°C for 1 hour to take the water out of them. Results showed that furnace dried samples have a smooth surface and are more resistive than the bubbled composites. The bubbled samples were most likely less resistive than samples with a smooth surface because during the bubble formation the CNTs are being pushed closer together causing percolation to happen sooner.

1. Introduction

Electrically conducting composite polymers are important because they give the best of both worlds when it comes to polymers and conducting materials. Conducting materials are typically very dense and brittle. Polymers can give a large range of properties such as improved impact resistance, be lightweight, and yet malleable when needed. The beauty of a conducting composite polymer is that one can achieve all these properties at the same time. This allows a whole surplus of applications like; replacing metal wires were weight must be optimized, the creation of self-regulating heaters, electric noses, chemical sensors, supercapacitors for charge storage, and electromagnetic shielding materials [1-9].

In this study, Acrylonitrile Butadiene Styrene (ABS) is the polymer used to make up the matrix of the composite. ABS is an amorphous polymer which is impact resistant. It is used to make Legos [10] and is often the polymer of choice for 3D printing [11]. ABS is made from three monomers (Acrylonitrile, Butadiene, and Styrene). It has a long linear backbone with branches of polybutadiene coming out the sides. ABS is a highly resistive polymer with a resistivity of 14-16x1015 ohm.cm3, meaning it is highly insulating and does not conduct electricity [12]. Carbon Nanotubes (CNTs) are the electrically

conducting material used in this study. CNTs are very conducting with a resistivity of 2.80x10-3 ohm.cm [13]. They are relatively small with a huge surface area and are made of only carbon atoms arranged into hybridized hexane rings [14]. There are two main types of CNTs: single-walled and multi-walled nanotubes[15]. For this study multi-walled CNTs were used.
When working with electrically conducting composites, it is important to determine their percolation threshold. Percolation occurs when there is enough of the conducting material present so that it can form a path, such that at that point the conductivity increases by several orders of magnitude [16]. The percolation threshold changes based on how the filler is dispersed in the matrix. The dispersion of the filler and the concomitant properties can be dramatically changed through different processing techniques [17]. In this study, the dissolution method is used to dissolve ABS with DMF and evenly disperse the CNT throughout the matrix.

2. Methods

2.1. Materials
Magnum ABS Resin supplied from Dow Chemical Company (Midland, Michigan), in the form of small pellets ranging in 5 mm in diameter and 2 mm in thickness. Short Multi-Wall Carbon Nanotubes, which had an outer diameter of 8-15 nm and a length of 0.5-2.0 μ m, were obtained from www.cheaptubes.com HPLC grade N, N-Dimethylformamide (DMF) was supplied by VWR Chemicals.

2.2. Fabrication
Varying concentrations of ABS/CNT composites were fabricated by first dissolving the ABS in DMF. (table 1) ABS was dissolved with the assistance of a speed mixer (Flack Tek, Inc., Model DAC 150.1 FVZ-K), which mixed the solution at 2000rpm for 1-minute intervals until all the ABS was completely dissolved. Then, the CNTs were added and mixed in using the speed mixer. The solution was allowed to dry in petri dishes inside the fume hood for ~10 days. Once composites were dry, they were diced finely and then compressed in a Compression Molder Struers Prontopress-2. The Compression Molder was heated to 170°C for 15 minutes and cooled for 8 minutes with a load of 10 kN. Compressed composites had a diameter of 3.17 cm and a radius of 1.585 cm. Some diced ABS/CNT composites were also dried in a furnace (Thermolyne model 47900) before compression. Samples were heated at 100, 120, or 150°C for 1 hour before compression. After compression samples were imaged using a Pent aView optical microscope.

Table 1. Carbon Nanotube Concentrations Used to Fabricate ABS/CNT Composites.

phr	wt %
0.000	0.000%
0.010	0.010%
0.250	0.249%
0.100	0.100%
0.750	0.744%
0.500	0.498%
1.000	0.990%
5.000	4.762%
10.000	9.091%

2.3. Impedance measurements
Composites were coated in silver using a Denton Vacuum Desk II Turbo Sputter Coater. Then, electrical measurements were taken using a Solartron 1260 Impedance Analyzer coupled with a 1296 Solartron Dielectric Interface (Farnborough, Hampshire, England). A two-probe test set up was used and measurements were taken at frequencies from 10^7-10^{-3} Hz at room temperature.

3. Results and discussion

3.1. Bubbled samples

Figure 1 shows compressed ABS/CNT composite samples that were not oven-dried. These samples had concentrations ranging from 0 to 10 phr. All of these composites have a foamed structure (figure 1 (a-g)). High concentrations such as 1 phr have phase separation in some areas where CNTs visually appeared to be absent or bubble size was dramatically different in size (figure 1 (h)).

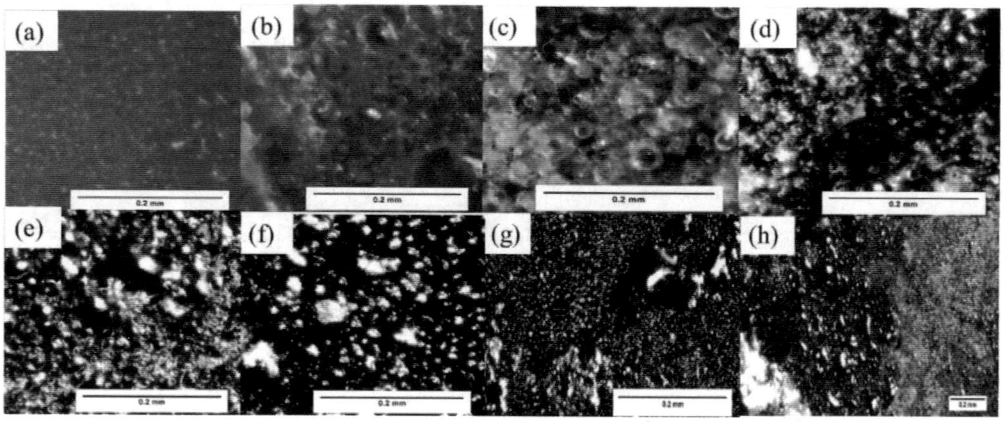

Figure 1. Macroscopic pictures of samples at different concentrations that all had a bubbling structure taken with the PentaView. (a) 0 phr (b) 0.01 phr (c) 0.1 phr (d) 0.5 phr (e) 1.0 phr (f) 5.0 phr (g) 10.0 phr (h)) 1 phr sample that shows two different regions. The left side has a higher concentration of CNTs and larger bubbles as compared to the right side.

Figure 2 shows the electrical measurements of the bubbled samples. Percolation happens somewhere between 1 and 5 phr for the bubbled samples. The standard deviation between samples of the same concentration was approximately the same for all concentrations. This is most likely due to the samples being inhomogeneous. There is a notably large variation in resistivity between samples around percolation, this is most likely because at percolation electrical pathways are starting to form and because this process is random, more pathways can form in some samples than in others.

IOP Conf. Series: Materials Science and Engineering **842** (2020) 012003 doi:10.1088/1757-899X/842/1/012003

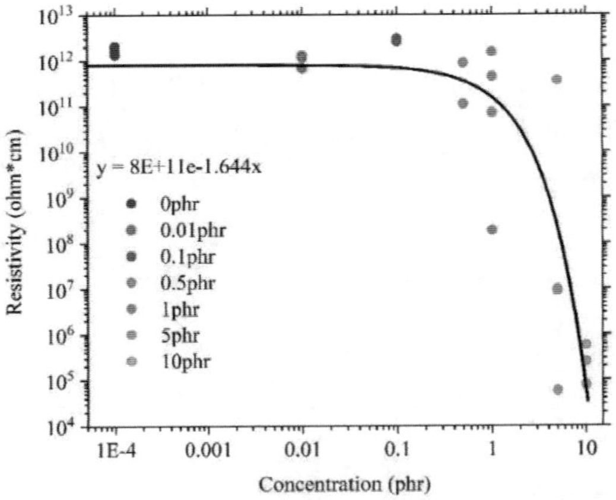

Figure 2. Resistivity vs CNT concentration in phr for ABS/CNT bubbled composites. Line of best fit is y= $8x10^{11}e^{-1.644x}$. Percolation happens around 1-5 phr. 0 phr was graphed as 0.0001 phr.

3.2. The cause of the bubbles in ABS/CNT composites

Bubbling was determined to be coming from water that entered the composites while they were air drying in the fume hood. This was determined by Fourier-transform infrared spectroscopy (FT-IR). Figure 3 shows FTIR spectra for as received ABS, ABS that was dissolved in DMF and dried in the fume hood, and compressed samples. The ABS that was dissolved in DMF has a very distinctive OH bond, which is found in water. The OH stretch is absent in the compressed ABS sample indicating that the water evaporated during compression which is most likely the source of the bubbles in the compressed samples. Thermogravimetric analysis (TGA) was also used to confirm that water was the cause of the bubbles.

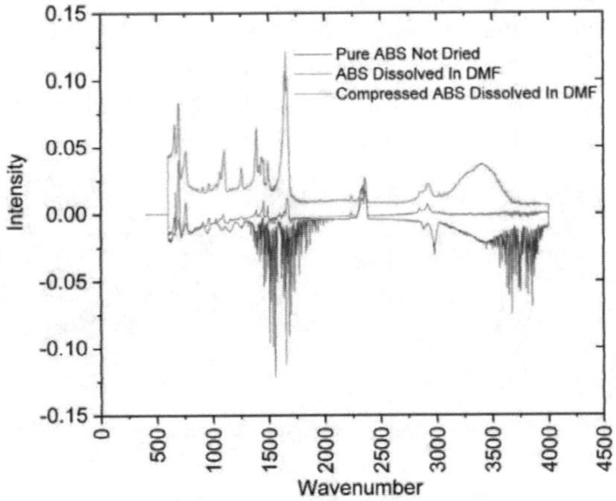

Figure 3. FT-IR of pure ABS, ABS dissolved in DMF and dried in the fume hood for 10 days, and ABS dissolved in DMF that had been compressed in the compression molder.

3.3. Oven drying samples before compression to eliminate the bubbling

Figure 4 shows PentaView images of diced ABS/CNT composites that were heated in a furnace at 100, 120, or 150°C for 1 hour before compression. Figure 4 (a, d, g) shows that samples that are heated at 100°C for 1 hour have the least amount of bubbles after compression. Although the 100°C treatment did not create smooth samples in ABS/CNT composite concentrations higher than 0.25 phr (figure 4 j, l, m). Samples without bubbles appeared smooth.

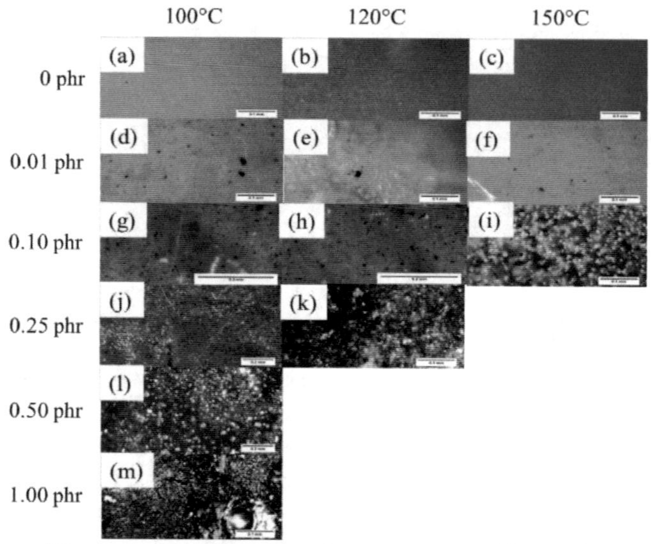

Figure 4. PentaView photos of samples that were treated in the furnace for 1 hour.

Figure 5 shows that heated samples were more resistive in comparison to their bubbled counterparts. The 0.01 phr sample that was heated at 100°C for 1 hour was 169.74% more resistive than the bubbled sample (figure 5 (a)) and the 0.1 phr heated sample was 87.28% more resistive than its bubbled counterpart(figure 5 (b)). This is most likely due to the bubbles pushing the CNTs closer together, causing percolation to happen sooner in the bubbled samples.

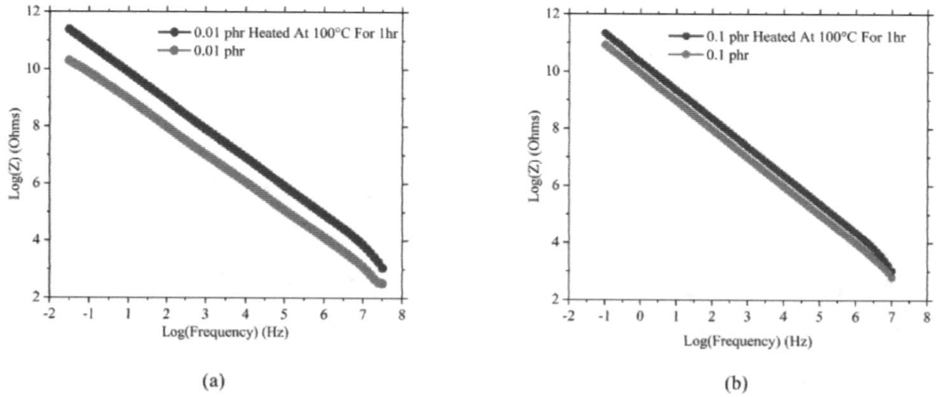

Figure 5. Graphs of Impedance (Z) vs Frequency data (a) 0.01 phr composites comparing heating in the furnace for 1 hour at 100°C with no treatment (b) 0.1 phr composites comparing heating in the furnace for 1 hour at 100°C with no treatment.

4. Conclusions

The ABS/ CNT composite polymers that had a foamed structure reached percolation between 1 and 5 phr. It was determined that the foaming structure came from water that had entered the samples during processing. Water was able to be removed from the samples before compression with the use of a furnace, this process had the most success when samples were heated to 100°C for 1 hour. Heating in the furnace only worked on lower concentrations of ABS/CNT. Concentrates higher than 0.5 phr plastically deformed in the furnace at 100°C. This is most likely due to the CNTs conducting heat through the samples causing the ABS to reach its glass transition point faster and at lower temperatures. This data shows how important processing is when making electrically conducting composite polymers. By simply heating the composites before compression the resistivity and microstructure was completely changed. This means that with all the same starting materials, one could easily manipulate the properties of the composite, possibly resulting in additional potential applications for the composite. More research into the processing of these composites is needed. The furnace heated samples need to be made with a larger range of concentrations so that they can be compared more directly to the foamed composites. Also, now that it is known that water can enter the composites when they are drying in the fume hood, the time that they are allowed to air dry needs to be more strictly controlled. In future studies, experimentation with drying samples in a vacuum oven might result in more reliable results across all concentrations of ABS/CNT composites. With the vacuum oven, there is a possibility of removing the water from the samples at lower temperatures so that the plastic deformation of the high concentrated composites may be avoided. Creating and testing more concentrations around percolation should be done in the future. This way a more accurate percolation concentration can be achieved.

5. References

[1] R. Strumpfler, J. Glatz-Reichenbach (1999). Conducting Polymer Composites. *Journal of Electroceramics 3[4]:* 329-346.

[2] J.N. Aneli, G.E. Zaikov, O.V Mukbaniani, C. Sirghiem, V.M. Lobo (2011). Physical Principles of the Conductivity of Electrical Conducting Polymer Composites: Review. *Journal of Characterization and Development of Novel Materials 3[3/4]: 139-162.*

[3] S.M. Mirvakili, I.W. Hunter (2018). Artificial Muscles: Mechanisms, Applications and Challenges. *Advanced Materials 3C[6].*

[4] Kai Wang, Haiping, Wu, Yuena Meng, Zhixiang Wei (2014). Conducting Polymer Nanowire Arrays for High Performance Supercapacitors. *Small 10[1]: 14-31.*

[5] W.E. Jones, J. Chiguma, E. Johnson, A. Pachamuthu, D. Santos (2010). Electrically and Thermally Conducting Nanocomposites for Electronic Applications. *Materials 3: 1478-1496.*

[6] A. Saboor, A.N. Khan, R. Jan, S. Sharif, M. Khan (2018). Mechanical, dielectric and EMI shielding response of styrene acrylonitrile, styrene acrylonitrile/polyaniline polymer blends upon incorporation of few layer graphene at low filler concentrations. *Jorunal of Polymer Research 25: 24.*

[7] GX Wang, A. Morrin, M. Li, N. Liu, XL Luo. (2018). Nanomaterials doped conducting polymers for electrochemical sensors and biosensors. *Journal of Materials Chemisstry B 6[25]: 4173-4190.*

[8] K.M. Vighnesha, S. Sandhya, D.N. Sangeetha, M. Selvakumar (2018). Synthesis and characterization o activated carbon/conducting polymer composites electgrode for supercapacitor applications. *Journal of Mateirals Scince: Mater Electronics 29: 914-921.*

[9] S. Ganguly, P. Bhawal, R. Ravindren, NC Das (2018). Polymer Nanocomposites for Electromagnetic Interference Shielding: A Review. *Journal of Nanoscience and Nanotechnology 18[11]: 7641-7669.*

[10] *WHAT LEGO® BRICKS ARE MADE FROM.* (2019). Retrieved from LEGO: https://www.lego.com/en-us/service/help/bricks-building/brick-facts/what-lego-bricks-are-made-from-408100000007855

[11] *ABS Filaments For Your 3D Printer.* (n.d.). Retrieved from Filaments.directory: https://www.filaments.directory/en/plastics/abs

[12] S. Dul, L. Fambri and A. Pegoretti (2018). Filaments Production and Fused Deposition Modeling of ABS/Carbon Nanotube Composites. *Nanomaterials 8[49]: nano8010049.*

[13] WY Wang, GH Luo, J. Luo (2009). Electrical Conductivity and Thermal Properties of Acrylonitrile-Butadiene-Styrene Filled with Multwall Carbon Nanotubes. *Polymer Engineering and Science 49[11]: 2144-2149.*

[14] Guillaume Herlem, F. P. (2019). Carbon Nanotubes: Synthesis, Characterization, and Applications in Drug Delivery Systems. In *Nanocarriers for Drug Delivery Nanoscience and Nanotechnology in Drug Delivery* (pp. 469-529).

[15] James P. Schaffer, T. H. (2011). *Principles and applications- Engineering Materials Vol 1 Georgia Institute of technology, Materials science & engineering*

[16] Runqing Ou, Siddhartha Gupta, Charles Aaron Parker, Rosario A. Gerhardt (2006). Fabrication and Electrical Conductivity of Poly(methyl methacrylate) (PMMA)/Carbon Black (CB) Composites: Comparison between an Ordered Carbon Black Nanowire-Like Segregated Structure and a Randomly Dispersed Carbon Black Nanostructure. *The Journal of Physical Chemistry, 110(45)*, 22365-22373.

[17] Sidhartha Gupta, Runqing Ou, Rosario A. Gerhardt (2006). Effect of the Fabrication Method on the Electrical Properties of Poly(acrylonitrile-co-butadiene-co-styrene)/Carbon Black Composites. *Journal of Electronic Materials*(2): 224-229.

Acknowledgments

Bryce Estell's summer internship at the Georgia Institute of Technology was funded by Prof. Gerhardt's Goizueta Faculty Chair. The authors acknowledge technical assistance from Morgan R. Watt.

IOP Conf. Series: Materials Science and Engineering **842** (2020) 012004 doi:10.1088/1757-899X/842/1/012004

Is Hygrothermal Aging of Construction Polymer Composites a Reversible Process?

Sotirios Grammatikos[1, a)], Styliani Papatzani[2, 3, b)] and Mark Evernden [4, c)]

[1]Group of Sustainable Composites, Department of Manufacturing and Civil Engineering, Norwegian University of Science and Technology, Gjøvik, Norway
[2]Hellenic Army Academy, Department of Mathematics and Engineering Sciences, Vari, Greece
[3]Hellenic Ministry of Culture, Directorate of Restoration of Medieval & Post-medieval Monuments, Athens, Greece
[4]Department of Architecture & Civil Engineering, University of Bath, Bath, United Kingdom

[a] sotirios.grammatikos@ntnu.no

Abstract. This paper presents the effects of wet/dry cycling loading on the moisture uptake behavior of a Fibre Reinforced Polymer (FRP) composite used in the civil engineering sector. FRP samples of various dimensions were cut from an 'off-the-shelf' pultruded flat sheet and conditioned in a cyclic hygrothermal environment. A series of 3 consecutive moisture absorption-desorption cycles lasting for 153 days were carried out to investigate the moisture uptake behavior of FRPs. The hygrothermal procedure consisted of immersion in 60°C distilled water until saturation and consecutively drying in a 60°C oven until equilibrium was reached. After the 1st desorption cycle, it was found that FRP samples lose a significant amount of mass due to chemical decomposition, the extent of which increases as wet/dry cyclic loading progresses. The effective mass loss leads to a subsequent significant increase in the rate of moisture uptake. Mechanical behavior of the FRPs aged at 40°C, 60°C and 80°C for 224 days is examined at both 'wet' and 'dry' states to reveal the reversible and irreversible effects of moisture uptake. It was revealed that the effects of 40°C hygrothermal aging on mechanical performance are reversible when moisture is removed.

1. Introduction

Pultruded FRPs (FRPs) are commonly used materials in the construction sector. The common cross section of a FRP consists of E-glass fibre reinforcement in a thermoset matrix [1]. FRPs are being employed as primary and secondary structural elements providing high resistance to extreme environmental conditions such as freeze-thaw, UV, etc. [2]. In addition to bridge engineering, civil FRPs [3] are used in a variety of other engineering applications including facades, off-shore and on-shore platforms, wind turbine blades, ladders and composite utility poles [4, 5], as well as reinforcement of concrete and steel structures [6].

In off-shore applications, the FRPs would be partly or fully submersed in water and subjected to wetting and drying cycles. It is therefore, essential that an experimental regime of simulated accelerated aging process is devised, assessing the behavior of FRPs during absorption and desorption cycles.

Thus far, the diffusion coefficients in the three principal directions have been determined for selectively aging conditions at 60°C through the application of one-dimensional Fickian theory. Mechanical, viscoelastic and physico-chemical analyses evaluated the moisture uptake of the FRPs and verified the decomposition taking place during the aging of FRP samples after 224 days of immersion [7, 8]. Elemental analysis has showed no chemical degradation incidents on the fibre reinforcement surfaces and infrared spectroscopy revealed superficial chemical alteration in the aging matrix. Therefore, it was reported that chemical decomposition, which is simply seen as mass loss of the initial dry mass of samples, stems from the 'weak' fibre sizing [1, 7]. Moreover, a sensing system based on Electrical Impedance Spectroscopy (EIS) has been developed to reveal 'true' moisture uptake characteristics of FRPs, independent of mass loss due to decomposition effects [9]. It is of particular interest though, to assess the effect cyclic hygrothermal loading has on FRPs and the possibility to reverse the effects of aging. The results are significant for the development of durable composites exposed in wetting and drying environments, while research can be extended to loading combinations including harsh environment and extreme loading requirements.

2. Background

When FRP samples are exposed to moist environments, moisture concentration increases with time and reaches a saturation point after an extended period of time which is always dependent on the exposure temperature, the type and thickness of the material. Generally, moisture absorption in polymer composites follows a Fickian diffusion trend [10]. However, Fickian theory is unable to describe the moisture diffusion process when significant mass loss due to chemical decomposition occurs simultaneously with moisture absorption [1]. Figure 1 depicts 3 moisture uptake vs. time curves that are representative of a) a classical Fickian diffusion three-stage curve with no mass loss, b) a diffusion curve affected by mass loss taking place, and, c) a 'true' diffusion curve when mass loss takes place without affecting the moisture uptake measurements.

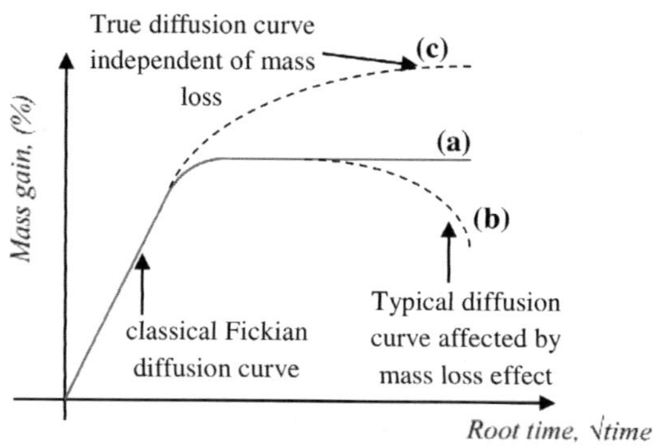

Figure 1. Representative plot of a) a classical Fickian diffusion three-stage curve with no mass loss, b) a diffusion curve affected by mass loss and c) a 'true' diffusion curve when mass loss takes place without affecting moisture uptake measurements for Mass gain (%) vs. $\sqrt{t}$

Water uptake monitoring conducted via gravimetric measurements often results into case 'b' curves of Figure 1. It is therefore interesting to study the effect of cyclic hygrothermal loadings on the water uptake characteristics of FRPs.

3. Experimental procedure

3.1. Materials

A 5-ply commercially available glass FRP pultruded profile [FS040.101.096A] was tested provided by 'Creative Pultrusions Inc., PA, USA'.

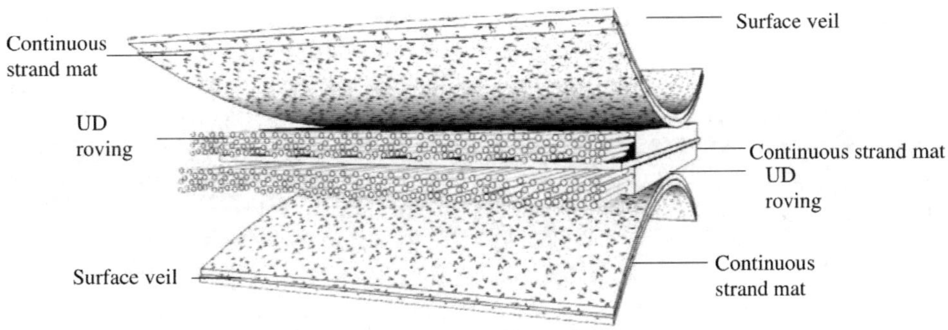

Figure 2. FRP profile schematic representation.

The nominal thickness of the profile was approximately 6.4 mm. The outer surfaces of the laminates are covered by a protecting and non-structural polyester layer (veil) which has the dual functions of retarding moisture ingress and protecting the FRP material from UV radiation. E-glass fibres served as the reinforcement and an isophthalic polyester resin formed the matrix of the composite. Figure 2 shows the structure of the tested FRP material which consisted of 3 continuous strand mats (CSM) and 2 unidirectional (UD) layers with 33.3% fibre and 54.5% fibre volume fractions, respectively.

3.2. Hygrothermal aging

Cyclic hygrothermal loadings were conducted on 40x40mm samples immersed in 60°C distilled water. All samples were dried in an over for 48hrs at 30°C to remove any moisture absorbed by the environment. Samples were subjected to a total of 3 absorption-desorption cycles. Moisture absorption cycles continued up until samples had gained 0.5% of mass compared to their initial dry state. Then samples were dried in an oven at 60°C until equilibrium, prior to water immersion for the next cycle. Gravimetric measurements were conducted using a digital analytical scale with ±0.001mg accuracy. Relative (*M(%)*) moisture uptake increase was determined using Eq. 1 (ASTM D5229), see figure 3:

$$M(\%) = \frac{M_t - M_0}{M_0} * 100\% \tag{1}$$

where M_t is the measured moisture mass at time t and M_o the initial dry mass (reference state). Moisture uptake measurements were conducted over a period of up to 224 days [1].

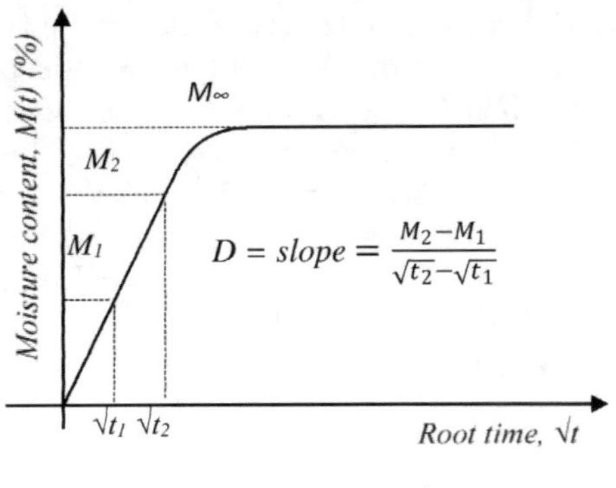

Figure 3. Diagram of a classical Fickian diffusion three-stage curve for M(t) vs. and the maximum absorbed moisture content M∞.

3.3. Mechanical testing

In order to assess the effects of hygrothermal cyclic loadings on the behaviour of the FRP material, 'Plate twist' testing was adopted. Plate twist is a standard in-plane shear mechanical test for polymer composites. During plate twisting, samples are subjected to minimal flexural loads which practically do not affect the structural performance of the material. For practicality reasons, it was assumed that samples are not influenced by plate twisting and therefore the same samples were used for both wet and dry shear tests. Plate twist shear tests were conducted according to the ISO 15310 standard for testing [11]. Samples were cut in 200x200x6.4mm dimensions and subjected to water immersion at 40°C, 60°C and 80°C for a period of 224 days. After aging, all samples were tested in shear in the wet state. Subsequently, samples were dried in an oven (drying temperature identical to hygrothermal aging temperature), until equilibrium was reached. After drying, samples were then tested in their dry state.

4. Results and discussion

Figure 4 displays a representative 3-cycle moisture absorption-desorption curve. As can be seen, it takes less days for the samples to reach 0.5% mass gain, during the 2nd and significantly less during the 3rd cycle loading.

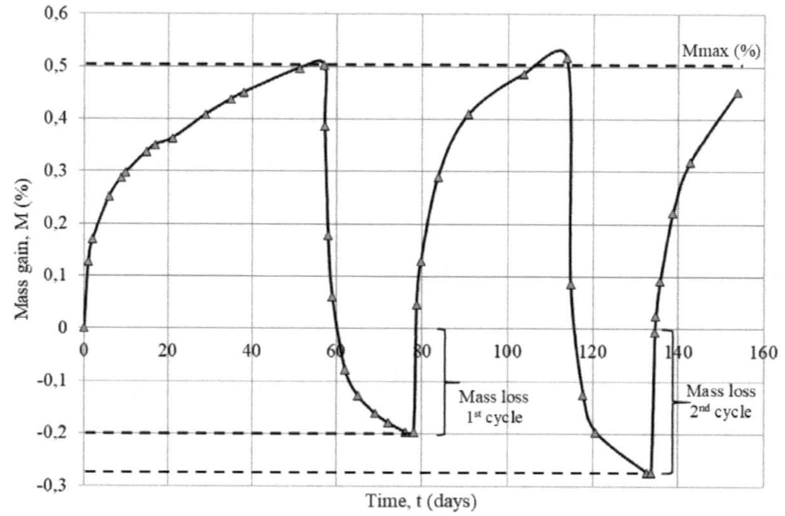

Figure 4. Moisture absorption and desorption curve for one sample hygrothermally aged at 60°C.

This effect is a result of significant mass loss that takes place during hygrothermal aging and is revealed during the drying phases of the material. Chemical decomposition leads to mass loss which practically allows for leachates to be washed out from the FRP structure. Chemical analysis of the aging medium has proved that leachates stem from the fibre/matrix interface paving the way for fibre/matrix interfacial failure [1, 7]. At the same time, hygrothermal aging along with wet/dry cyclic loading promote matrix cracking. The combination of interfacial failure and matrix-cracking leads to intrinsically more space or 'reservoir' for water molecules to penetrate altering the moisture uptake behavior of the material. Fibre/matrix interface failure, matrix-cracking and increasingly higher amount of moisture penetrating the structure are expected to promote a gradual deterioration of the FRP's structural performance. It could be postulated that decomposition may cease when all the active compounds are totally consumed.

With respect to mechanical performance, figure 5 illustrates the 'wet' and 'dry' shear modulus of the studied FRP after 224 days of hygrothermal aging in comparison with the reference (unaged) dry state. As can be seen, hygrothermal aging induces detrimental effects to the material when tested at its wet state. However, this is not the case for aged samples tested at their dry state. It was revealed that, samples aged at 40°C for 224 days, recover completely (100%) their performance when tested at their dry state.

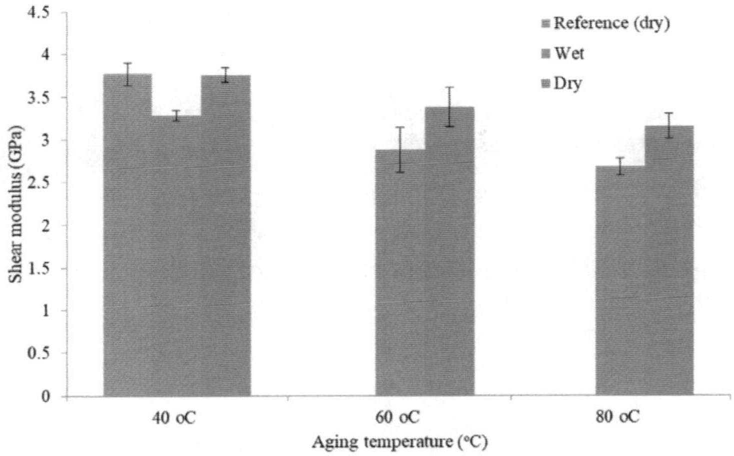

Figure 5. Shear modulus as a function of aging temperature for wet and dry samples after 224 days of aging.

In this respect, it becomes apparent that hygrothermal aging and mass loss due to decomposition become detrimental to the structural performance of FRPs at their wet state and less significant in the absence of moisture.

5. Conclusions

The effects of wet/dry cycling loading on the moisture uptake behavior of a commercially available pultruded polymer composite were investigated. The cyclic hygrothermal loadings demonstrate that the material losses mass from the initial dry mass due to chemical decomposition, which has an effect on the rate of moisture diffusion. Interestingly enough, the mechanical testing of wet and dry samples revealed that material decomposition is to a certain extent reversible when the material returns to its initial dry state.

6. References

[1] Grammatikos, S.A., et al., Moisture uptake characteristics of a pultruded fibre reinforced polymer flat sheet subjected to hot/wet aging. Polymer Degradation and Stability, 2015. 121: pp. 407-419.

[2] Pochiraju, K.V., G.P. Tandon, and G.A. Schoeppner, Long-Term Durability of Polymeric Matrix Composites, Springer, 2011.

[3] Bakis, C., et al., Fiber-Reinforced Polymer Composites for Construction—State-of-the-Art Review. Journal of Composites for Construction, 6(2), (2002) pp. 73-87.

[4] Shao, Y. and S. Kouadio, Durability of fiberglass composite sheet piles in water. Journal of Composites for Construction, 6(4), (2002) pp. 280-287.

[5] The New and Improved PultexÒ Pultrusion Design Manual, C.P. Inc., Editor. Alum Bank, PA, (2014) p. 255.

[6] Linghoff, D., R. Haghani, and M. Al-Emrani, Carbon-fibre composites for strengthening steel structures. Thin-Walled Structures, 47(10), (2009) pp. 1048-1058.

[7] Grammatikos, S.A., et al., On the response to hygrothermal aging of pultruded FRPs used in the civil engineering sector. Materials & Design, 96, (2016) pp. 283-295.

[8] Ullah, Z., et al., Multi-scale computational homogenisation to predict the long-term durability of composite structures. Computers and Structures, 181, (2017) pp. 21-31.

[9] Grammatikos, S.A., et al., Impedance spectroscopy as a tool for moisture uptake monitoring in construction composites during service. Composites Part A: Applied Science and Manufacturing, 2018. 105: pp. 108-117.

[10] Surathi, P., et al., Hygrothermal Effects on Durability and Moisture Kinetics of Fiber-reinforced Polymer Composites. 2006: Department of Structural Engineering, University of California, San Diego.
[11] Grammatikos, S.A., et al., Thermal cycling effects on the durability of a pultruded GFRP material for off-shore civil engineering structures. Composite Structures, 2016. 153: pp. 297-310.

Acknowledgments
The EPSRC project DURACOMP (Providing Confidence in Durable Composites, EP/K026925/1) is greatly acknowledged for financial support.

IOP Conf. Series: Materials Science and Engineering **842** (2020) 012005 doi:10.1088/1757-899X/842/1/012005

Abrasive Size and Fiber Direction Effects on the Wear Behaviour of CFRP Composite under Rotary Abrasion

N Anandan, M Ramulu*

University of Washington, Seattle, USA

*E-mail: ramulum@uw.edu

Abstract. Rotary wear tests were performed on unidirectional graphite fiber reinforced epoxy to study the effect of fiber orientation and abrasive particle size on the wear resistance and mechanism of the graphite/epoxy polymer composite. The tests were performed at a constant load of 1250 grams for 10000 cycles with abrasive particle sizes of 15 and 55 μm using Taber rotary platform abrasion tester. The surface topology and material removed along the wear track were obtained for a range of fiber orientations. The surface roughness was found to decrease as the fiber orientation increased from 0° to 90°. Negative skewness indicated the presence of numerous valleys on the surface. These variations in the surface roughness parameters were more pronounced for higher particle size of 55 μm. Maximum wear occurred at 135°. Further, the wear mechanism and type of damage were identified through SEM micrographs. Distinct damage could be identified at 0°, 45° and 90°.

1. Introduction

The innovations in the development of polymer composites has allowed its extensive application in the aerospace and automobile industries. These composites have tailorable properties with high strength to weight ratio, excellent fatigue and corrosion resistance in comparison to metal alloys[1,2]. Additionally, when polymer composites are being used in sliding wear applications, such as, ball joints in the car chassis, the wear performance should be thoroughly characterized [3]. Although the wear resistance of polymer composites have been studied extensively as a function of volume fraction of the fiber, type of fiber and matrix material[4–6], only a few studies have reported the effect of fiber orientation[7,8]. A thorough knowledge of wear behaviour of polymer composites is critical to further extend its use to other applications.

The degree of abrasiveness of the fibers, thermal stability, thermal conductivity of the matrix and coefficient of friction play an important role in characterizing the wear behaviour[9]. Polymer composites reinforced with short fibers such as carbon and glass fibers were reported to significantly improve the wear resistance. The abrasive wear debri from the carbon fiber was observed to have smoothening effect on the roughness of the counterface. The thermoset and thermoplastic polymer matrices have distinct modes of failure, thermoset polymers do not soften due to frictional heat and the generated heat induces chemical degradation and wear on the surface. These polymer composites do not yield due to softening and catastrophic failure[10]. Solid lubricants are added to these matrices to form a transfer film between contacting surfaces to prevent the chemical degradation. Specifically, for epoxy matrix the addition of graphene resulted in good wear life and lower value of coefficient of friction.

Owing to the anisotropic behavior of composites with continuous fiber, a few studies have reported the wear behavior as a function of fiber orientation [7,8]. The sliding wear test was performed in

IOP Conf. Series: Materials Science and Engineering **842** (2020) 012005 doi:10.1088/1757-899X/842/1/012005

directions normal, anti-parallel(90 degrees) and parallel(0 degrees) to the fibers. The wear rate was reported to be the lowest in the normal direction and inconclusive between the parallel and anti-parallel directions. This warrants for additional investigation to identify the wear rate at a range of fiber contact angles. This study investigates the wear rate of unidirectional graphite/epoxy laminate with 0 to 180° fiber orientations. This is accomplished using Taber rotary abrader under a constant load of 1250 grams for an abrasive particle size of 55 and 15 μm. As the platform rotates, the fiber contact angle changes from 0 to 180 degrees. The depth of material removed, surface roughness was studied for a range of fiber orientations. Additionally, the wear mechanism was identified through the SEM analysis of the wear tracks.

2. Materials and Experiment methods

The material used in this study was a graphite/epoxy unidirectional laminate with 55% fiber volume fraction, IM-6 graphite fibers embedded in 3501-6 epoxy resin. Unidirectional specimen of 100mm x 100mm x 2mm were fabricated using diamond saw from a panel of dimension, 1m x 0.5m x 0.002m. Specimen were tested on Taber rotary platform abrasion tester, shown in Figure 1 a at 72 rpm under a constant load of 1250 grams for 10000 cycles and the wear debri was removed continuously through suction. The experiments were performed according to ASTM D4060.

Sand paper strips with an average particle size of 15 and 55 microns were used as the counter face on the Tabor wheel. Since counterface plays a dominant role in determining the wear, the sand paper was replaced every 500 cycles to minimize the variations arising from blunting of the sharp edges. As the abrader completes one full rotation, the abrasive particles encounter fibers oriented from 0 through 180. For the all the experiments the abrasive wheel was positioned at 0 degree fiber orientation at the beginning of the wear test. Surface roughness measurements were performed using Mahr surface profilometer and Keyence VR300. The 2 D surface roughness was quantified maximum peak to valley height roughness, R_t, and ten point average roughness, R_z and maximum profile valley depth, z over a profile length L are defined by equations 1 through 2,

$$R_t = |\max(z) - \min(z)| \qquad (1)$$

$$R_z = \frac{1}{5} * [\sum_{i=1}^{5} \max(z_i) + \sum_{i=1}^{5} \min(z_i)] \qquad (2)$$

The 3D area height parameters, S_a, and S_z, were defined by equations 3 and 4,

$$S_a = \frac{1}{A} * (\iint |Z(x,y)|\ dxdy \qquad (3)$$

$$S_z = S_p + S_v \qquad (4)$$

Where, $Z(x, y)$ is the profile height as a function of x and y and, S_p is the largest peak height, and S_v is the deepest valley depth.

Wear resistance was assessed by the depth of material removed for a range of fiber orientations. The average depth was calculated by measuring the concave volume of material removed over a specific area using Keyence VR300. Additionally, scanning electron microscopy was utilized to evaluate the microscopic behavior of fibers and matrix along the wear track.

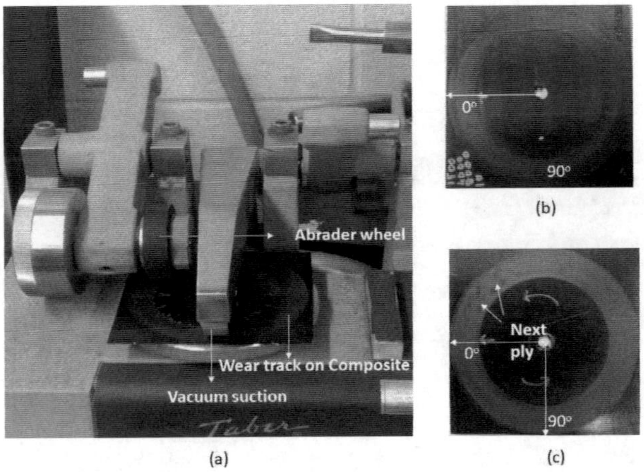

Figure 1. a) Experiment set up, Polymer composite mounted on the Taber rotary platform abrasion tester, b) specimen showing wear track with 15 µm abrader after 1000 cycles, c) specimen showing wear track with 55 µm abrader after 10000 cycles.

3. Results and Discussion

With the progression of the rotary wear test, the next ply was visible at certain fiber orientations, shown in Figure 1 b and c. Clearly indicating that the wear rate is dependent on the fiber orientations and particle size. Further, it was discernible that the surface profiles varied with the fiber orientation, shown in Figure 2. The filtered surface profiles showed the presence of deep valleys for 55 µm and appeared smoother for 15 µm abrasive particles. Additionally, the surfaces showed a higher deviation from the nominal surface at 0° and 45° in comparison to 90°. The average surface roughness parameters with standard deviations are plotted in Figure 2 d and e.

For 55 µm, the R_z was found to decrease from 6.2 to 4.4 µm and R_t decreased from 7.6 to 5.3 µm from 0° to 90°. Although similar trends were observed for 15 µm abrasive particle size, the variations were more pronounced for higher particle size. Similarly, the skewness and kurtosis of the wear track were analysed to study the asymmetric spread of peaks and valleys on the surface. The skewness was found to be negative for all the fiber orientations owing to the presence multitude valleys on the surface, shown in Figure 2 e, the features below the nominal surface are shown in blue. In tribology, negatively skewed surface is considered good for lubrication owing the valleys trapping the lubricants. The presence of valleys on the surface can be attributed to the epoxy eroding at a higher rate and increasing the interfacial gap between the fibers. Further affirmed through the observation of area surface profiles, Figure 3, with the red peaks aligned along the fiber direction and the blue valleys dispersed around the peaks. The skewness values were found to be more negative at lower particle size, this can be attributed to the higher density of particles randomly oriented on the abrader resulting in higher number of valleys in comparison to higher particle size. Similarly, kurtosis of surface profile is a measure of the sharpness of peaks. The kurtosis was observed to increase with the skewness, indicating that the material was removed by the action of random abrasive particles. The increase in kurtosis indicated the presence of sharper peaks for 15 µm particle size.

Further, the volume of material removed over a specific area was measured to calculate the depth of material removed. The variations in the amount of material removed as a function of fiber direction is shown in Figure 3 a. The variations in the depth were observed to be more pronounced, as expected for higher particle size. For 55 µm, the erosion was found to maximize at 45 and 135°, with a maximum of 180 µm at 135°. However, at 0 and 90° the amount of material removed was observed to be the same, concurring with the observations made by [7]. While, for 15 µm, the maximum difference in depth was found to be within 10 µm. The wear mechanism was investigated to understand the differences in the amount of material removed along the wear track.

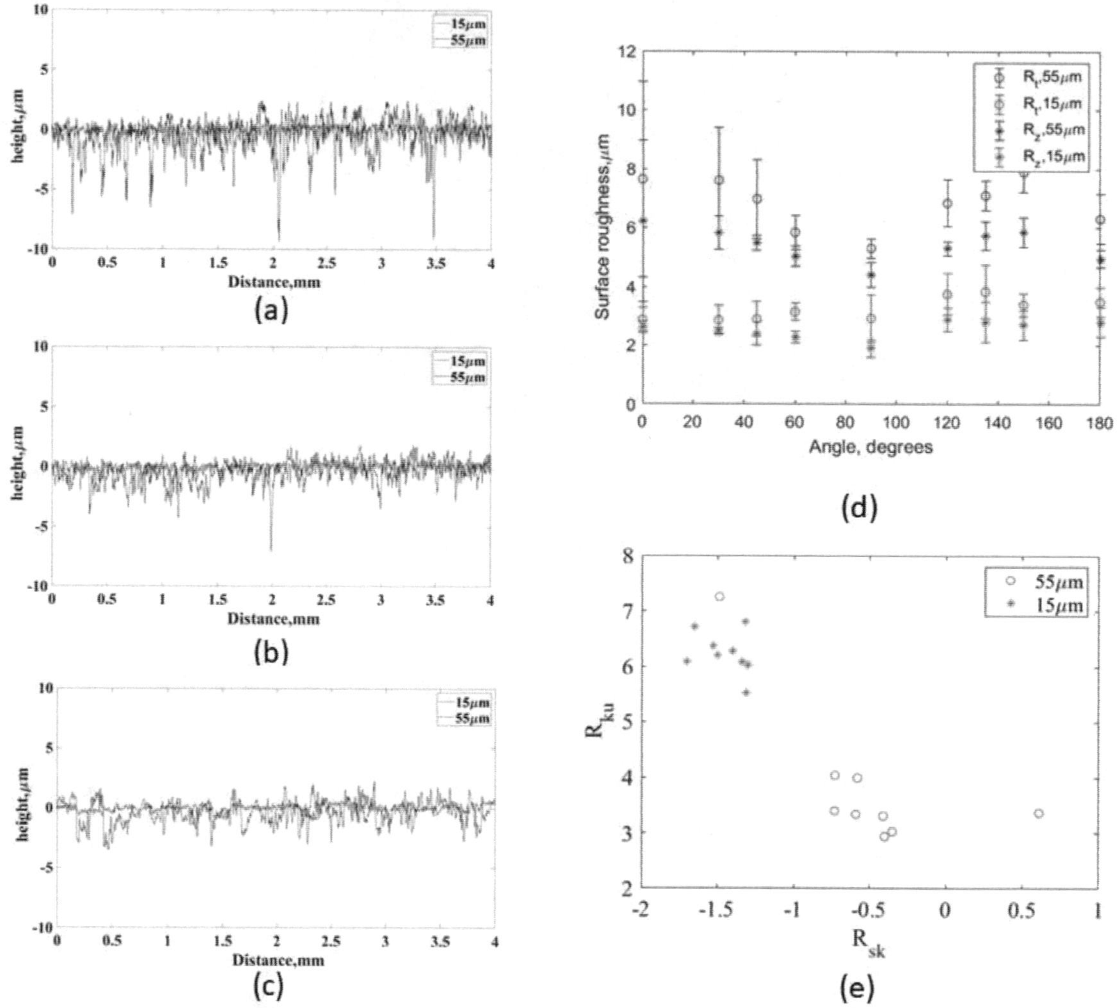

Figure 2. 2D surface roughness, a) Profiles of the wear track along the radial direction after 10000 cycles at, a) 0°, b) 50°, c) 90°, d) R_t, R_z as a function of fiber orientation, e) skewness vs kurtosis for 55 μm and 15 μm particles.

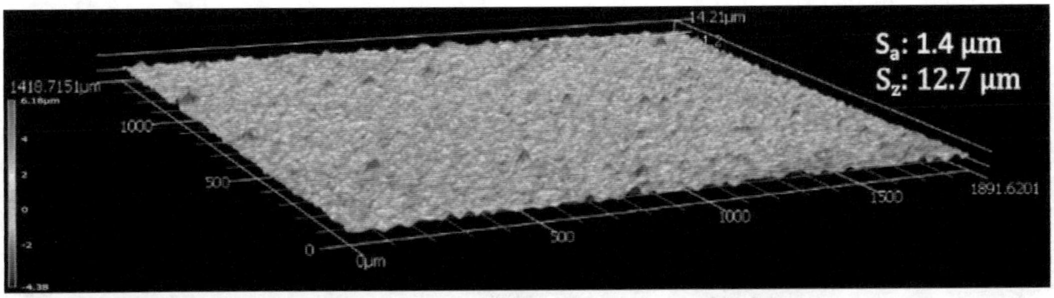

(a)

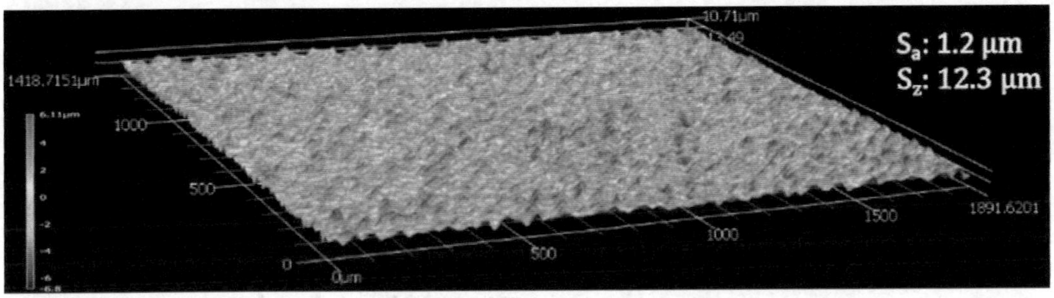

(b)

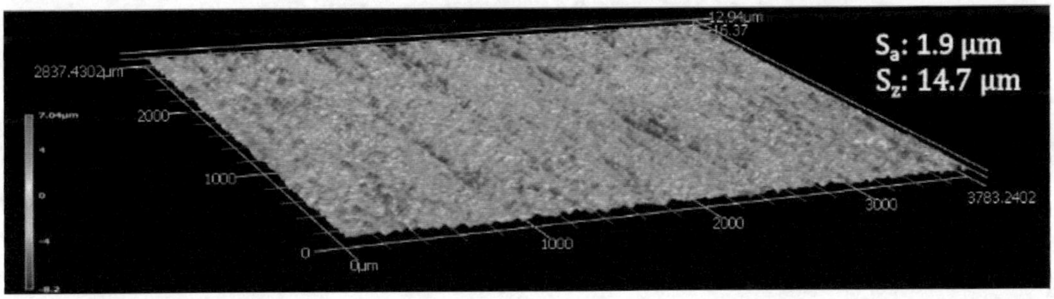

(c)

Figure 3. Area profiles at a) 0°, b) 45°, c) 90°, average S_a and S_z values are given on the top right, d) depth of material removed as a function of fiber orientation.

The SEM micrographs of the wear tracks showed broken fibers and epoxy debri dispersed along the wear track. Upon observation of damages to the fibers and the matrix, distinct wear mechanisms can be identified at different fiber orientations. Figure 4 a and b show the wear track created by 55 µm abrasive particles at 0°. Extensive fiber crushing, region A and fiber pull out, point B were found along the wear track. When the abrasive particles impinge along the fiber direction, the fibers may buckle and fracture, shown in region C. In addition, the epoxy matrix appeared to have failed by brittle fracture.

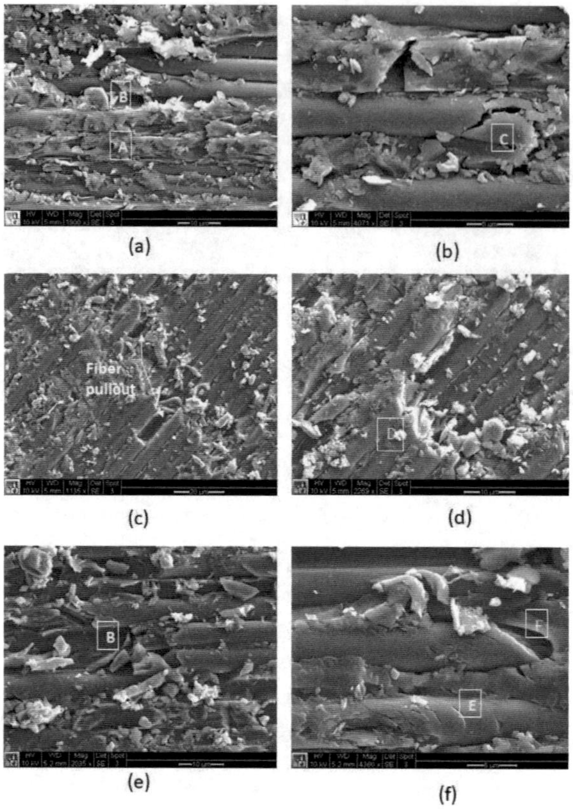

Figure 4. Wear track with 55 microns abrasive particles at fiber orientations of a, b) 0°, c, d) 45°, e, f) 90°.

Figure 4 c and d show the wear track at 45°. At this orientation, fiber cutting was observed to be dominant, shown in region D. Micro-cutting of fibers leads to the highest amount of wear in polymer composites and could be attributed to observing high wear at this angle. Large chunks of fibers are removed in addition to matrix cracking and fracture, contributing to the increased wear rate and visibility of the next ply in the region. At 90°, when the abrasive particles impinge normal to the fibers, a bending load is imparted to the fibers. In addition, the repeated impact resulted in the formation of microcracks, region E along the circumference of the fibers. The microfatigue resulted in crack growth and fiber fracture, region F.

4. Conclusion

The effect of fiber orientation on the wear resistance and mechanism was investigated in this study. The variation in surface profiles of the wear tracks were quantified interms of roughness parameters,

- The R_z and R_t were found to decrease as the fiber contact angle increased from 0° to 90°.
- A multitude of valleys were observed on the surface resulting in negative skewness of the surface. The skewness and kurtosis was found to increase with the decreasing particle size. This can be attributed to the higher density of particles on the surface.
- For 55 µm, the depth was found to maximize at 45 and 135°, with a maximum of 180 µm at 135°. A significant difference in the amount of material removed could not be observed between 0 and 90°.
- Distinct wear scars could be identified at 0°, 45° and 90°. Extensive fiber and matrix crushing was observed at 0°. While, microcutting of fibers was the dominant mechanism at 45°. Owing to the bending load on the fibers at 90°, fiber bending and fracture were observed to be dominant.

5. Acknowledgements
The authors sincerely acknowledge the Boeing-Pennell Professorship funding and Department of Mechanical Engineering support during this investigation.

6. Reference
[1] R.C. Alderliesten, R. Benedictus, Fiber/metal composite technology for future primary aircraft structures, in: J. Aircr., 2008: pp. 1182–1189.
[2] J.N. Coleman, U. Khan, W.J. Blau, Y.K. Gun'ko, Small but strong: A review of the mechanical properties of carbon nanotube-polymer composites, Carbon N. Y. 44 (2006) 1624–1652.
[3] A.M. HÄGER, M. Davies, Short-Fibre Reinforced, High-Temperature Resistant Polymers for a Wide Field of Tribological Applications, in: Compos. Mater. Ser., 1993: pp. 107–157.
[4] K. Friedrich, Z. Lu, A.M. Häger, Overview on polymer composites for friction and wear application, Theor. Appl. Fract. Mech. 19 (1993) 1–11.
[5] K. Friedrich, Z. Zhang, A.K. Schlarb, Effects of various fillers on the sliding wear of polymer composites, Compos. Sci. Technol. 65 (2005) 2329–2343.
[6] T. Tsukizoe, N. Ohmae, Wear performance of unidirectionally oriented carbon-fibre-reinforced plastics, Tribol. Int. 8 (1975) 171–175.
[7] J. Bijwe, R. Rattan, M. Fahim, Abrasive wear performance of carbon fabric reinforced polyetherimide composites: Influence of content and orientation of fabric, Tribol. Int. 40 (2007) 844–854.
[8] M. Cirino, K. Friedrich, R.B. Pipes, The effect of fiber orientation on the abrasive wear behavior of polymer composite materials, Wear. 121 (1988) 127–141.
[9] N.A.M. Tahir, M.F. Bin Abdollah, N. Tamaldin, H. Amiruddin, M.R. Bin Mohamad Zin, A brief review on the wear mechanisms and interfaces of carbon based materials, Compos. Interfaces. 25 (2018) 491–513.
[10] Failure Analysis and Prevention, ASM International, 2018.

IOP Conf. Series: Materials Science and Engineering **842** (2020) 012006 doi:10.1088/1757-899X/842/1/012006

Recycled Wood Plastic Composites as Floor Panels for Ship Containers

Sotirios Grammatikos[1, a)], Harald Bertin Jøsendal[1,b)], Styliani Papatzani[2, 3, c)], Orsa Boura[1,d)], Mohammad Hajmohammadian Baghban[1, d)]

[1]Group of Sustainable Composites, Department of Manufacturing and Civil Engineering, Norwegian University of Science and Technology, Gjøvik, Norway
[2]Hellenic Army Academy, Department of Mathematics and Engineering Sciences, Vari, Greece
[3]Hellenic Ministry of Culture, Directorate of Restoration of Medieval & Post-medieval Monuments, Athens, Greece

[a] sotirios.grammatikos@ntnu.no

Abstract. Within the framework of this study, we aim at the replacement of plywood as ship containers floor material, with a lighter and friendlier-to-rainforests composite material. Wood fibre-reinforced thermoPlastic Composites (WPC) were developed using double-screw extrusion, consisting of recycled waste wood fibres reinforcing recycled High Density PolyEthylene (HDPE) matrix. This shift to WPCs has the dual positive effect of containers freight reduction as well as elimination of the use of natural resources, radically contributing to climate change measures. Following international shipping standards for container flooring, high quality recycled WCPs were developed. HDPE was found to provide acceptable structural performance. To that respect, WPC blends of HDPE matrix reinforced with 30%, 40% and 50%wt wood fibre flour, were fabricated and subjected to 3 point-bending mechanical testing. It was found that the studied WPCs reveal mechanical performance adequate to qualify as replacements of plywood timber flooring in ship containers.

1. Introduction

The increasing demand for greener transportation imposes the need for new environmental-friendly engineering materials to reduce carbon footprint and Green-House Emissions (GHE) [1, 2]. In total 72 million containers operate in off-shore and in-shore missions, a third of which is used for sea transportation [3]. Currently, all shipping containers come with a standard timber flooring based on plywood sourced from rainforests. Plywood's extensive use has led to the gradual disappearance of animal species as well as to the elimination of forests efficiency to capture CO_2 emissions [3].
At the same time, the majority of plastics and other polymer composites used in industries such as transportation and energy, end up in landfill or are being used for energy recovery mainly via incineration (according to The Incineration of Waste Directive, 2000/76/EC) [4]. A significant amount of wastes stemming from packaging or polymer composites ends up in landfill around Europe or other non-EU countries (EC Waste shipment regulations (259/93/EEC)) [5].
Together with plastics and other waste issues within the EU, around 700 Mt of bio-based waste mass is produced annually of which at present, only a small percentage is recovered or reused for economic benefit [6].

Therefore, it becomes apparent, that a holistic solution to preserve plywood and at the same time promote recycling is of primary importance. In this study, we aim at the development of Wood fibre-reinforced thermoPlastic Composites (WPC) as replacement of plywood as floor material of ship containers. WPC is a lighter and friendlier to the precious rainforests material replacement. This shift to WPCs as container floors has the dual positive effect of significantly reducing the freight of containers as well as eliminating the use of natural resources. This radically contributes to the environmental cost and paves the way towards circular economy. WPC were developed using double-screw extrusion, consisting of recycled waste wood fibres reinforcing recycled polyethylene (PE) [7, 8]. Following the tight international shipping standards for container flooring with respect to mechanical and hygrothermal performance, high quality lightweight recycled WCPs were developed. WPC blends of HDPE matrix reinforced with 30%, 40% and 50%wt wood fibre flour, were fabricated and subjected to 3 point-bending mechanical testing. It was found that the studied WPCs reveal acceptable mechanical properties and can qualify as candidate replacements of plywood timber flooring in ship containers.

2. Experimental procedure

2.1. Manufacturing

WPC were produced by a simple mixture of wood powder with thermoplastic polymers (figure 1). Before the mixing process, all raw materials were dried out in an oven at 60oC for 24hrs to remove moisture absorbed by the environment. Then, wood powder and the polymer matrix were mechanically mixed in their dry state (figure 2).

Figure 1. Raw materials employed for the production of WPC.

IOP Conf. Series: Materials Science and Engineering **842** (2020) 012006 doi:10.1088/1757-899X/842/1/012006

Figure 2. Raw materials employed for the production of WPC.

This mixture of figure 2 was then heated and placed within the extruder for extrusion into different profile shapes.

2.2. Mechanical testing

Three-point bending was employed to assess the mechanical performance of the developed WPC according to the ISO 178. Three-point bending was performed on a Zwick Z2.5 testing machine. All mechanical testing was conducted at room temperature conditions. Five standard specimens were fabricated and tested for each wood powder fraction (30%, 40% and 50%wt). Dogbone samples were prepared according to ISO A1 in dimensions of 50x28x305 mm.

3. Results and discussion

Figure 3 displays the flexural strength whereas figure 4 the flexural modulus of the fabricated WPCs, for the 3 different wood-powder volume fractions. As can be seen, flexural strength exhibits an increasing tendency with wood powder content increase from 30%wt to 50%wt. However, figure 4 demonstrates that flexural modulus increases significantly with the increase in wood powder content.

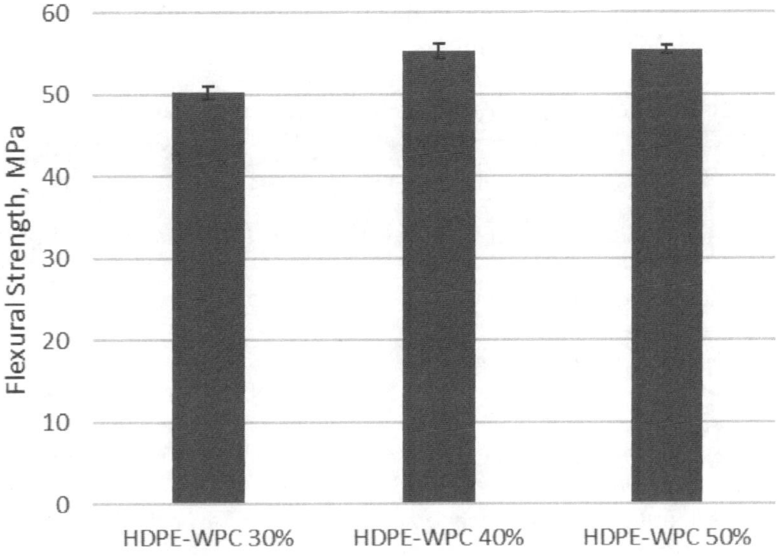

Figure 3. Flexural strength of the tested WPCs.

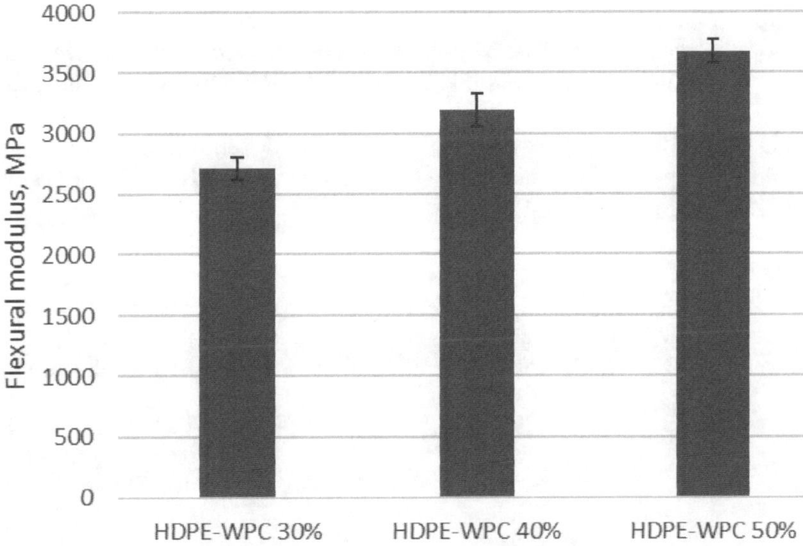

Figure 4. Flexural modulus of tested WPCs.

In order for the developed WPCs to qualify as a floor panel material for ship containers, they have to surpass the threshold of 65.9 MPa in flexural strength (according to the IICL TB 001 Performance standard for Structural Container Floor Panels) which is 10% higher that the WPC modified with 50%wt wood powder. This implies the fact that based on mechanical performance, the proposed WPCs are a promising sustainable material solution for the replacement of rainforest plywood in ship containers.

4. Conclusions

This study investigated the development of Wood Plastic Composites for potential replacement of plywood as ship containers floor material. WPCs were developed using double-screw extrusion, consisting of recycled waste wood fibres reinforcing recycled polyethylene (PE) matrix. WPC blends of HDPE matrix reinforced with 30%, 40% and 50%wt wood fibre flour, were fabricated and subjected to 3 point-bending mechanical testing. It was found that the studied WPCs reveal mechanical performance adequate to qualify as replacements of plywood timber flooring in ship containers. This replacement contributes to containers freight reduction (less CO_2), minimization of the use of fossil fuels and recycling, radically contributing to climate change measures for future circular economy.

5. References

[1] European Commission, Guidelines on financial incentives for clean and energy efficient vehicles, SWD, 27, 2013.

[2] European Commission, Communication: Integrating maritime transport emissions in the EU's greenhouse gas reduction policies, COM 479, 2013.

[3] Delpizzo, Rich, Haifeng Wang, and Andrew Panek. Sustainability in the Maritime Industry: A Collection of Relevant Papers. New York: Society of Naval Architects and Marine Engineers, Internet resource, 2011.

[4] European Commission, Directive 2000/76/EC of the European Parliament and of the Council on the incineration of waste, 2008.

[5] European Commission, Report from the Commission to the European Parliament and the Council on the implementation of regulation (EC) no 1013/2006 on shipments of waste, 2015.

[6] N. Reynolds, M. Pharaoh, An introduction to composites recycling, in: V. Goodship (Ed.), Management Recycling and Reuse of Waste Composites, Woodhead Publishing, (2010) pp. 3–19.

[7] M.C.S. Ribeiro, A.C. Meira-Castro, F.G. Silva, J. Santos, J.P. Meixedo, A. Fiúza, Reuse assessment of thermoset composite wastes as aggregate and filler replacement for concrete-polymer composite materials: a case study regarding GFRP pultrusion wastes, Resour. Conserv. Recycl., 2015.

[8] A. Yazdanbakhsh LC. Bank, C. Chen, Use of recycled FRP reinforcing bar in concrete as coarse aggregate and its impact on the mechanical properties of concrete, Construction and Building Materials 121, (2016) pp. 278–284.

Acknowledgments

This work was funded by the Regional Research Fund of Inland (Regionale Forskningsfond Innlandet - RFF Innladet - Forprosjekt WPC [Project reference: 280223]), Oppland County, Norway. Authors would like to acknowledge Norner Research AS and Woodrock Group AS for the support.

IOP Conf. Series: Materials Science and Engineering **842** (2020) 012007 doi:10.1088/1757-899X/842/1/012007

Development and Characterization of Alumina-Based Membranes for Water Purification

B A Ahmed*, F Ahmed, K Ali and J B Shabbir

Department of Mechanical Engineering, College of Electrical and Mechanical Engineering, National University of Sciences and Technology (NUST), Islamabad, Pakistan, 46000

*banjumahmed@gmail.com

Abstract. High energy ball milling (HEBM) and pressure less sintering technique were performed to achieve a homogenous powder mixture and porous ceramic membranes of α-Al_2O_3/2%MWCNTs. A uniaxial powder compaction press was used to compress the powder mixture into circular green pellets (27 mm in diameter, 3-4 mm thickness) at three different pressure values of 100 MPa, 250 MPa and 400 MPa, followed by sintering at three different temperature values of 1200 °C, 1250 °C and 1300 °C under argon gas atmosphere. The effect of process parameters on the pore distribution and phase change was characterized by employing Field Emission Scanning Electron Microscopy (FESEM) and X-Ray Diffraction (XRD) respectively. Process parameters were found to have a strong influence on the characteristics of the developed membranes like diametrical strength, water permeability and porosity. Water flow rates varied from 950 l/m^2.hr to 42 $l/m^2 \cdot hr$, porosity decreased from 48.57% to 28.87% and diametrical strength increased from 0.95 MPa to 11.06 MPa as compaction pressure was increased from 100 MPa to 400 MPa and sintering temperature was increased from 1200 °C to 1300 °C.

1. Introduction

Human activities are the major cause of water pollution around the globe[1]. In agricultural and industrial zones, fertilizers are observed to react in the soil to produce heavy metals as by products which further accrue in the soil and water[2].The main pollutants in these waste are nitrates, nitrites, cations, anions such as Ag^+, Na^+, K^+, Mg^{2+}, Ca^{2+}, Cl^-, HCO_3- and toxic metals like iron mercury, cadmium, lead, chromium, arsenic, nickel, cobalt, copper, zinc and magnesium[3, 4].

Adsorption based filtration technique has cemented its status as one of the economic and efficient way for removing heavy metals from wastewater. Many reported studies focus on the use of low-cost adsorbents for removing heavy metal ions from waste waters. Bhattacharyya and Gupta used kaolinite and montmorillonite to remove heavy metals from waste water[5]. Investigation have also been conducted for possibility of heavy metal removal through industrial by-products such as lignin, diatomite, clino-pyrrhotite, lignite, aragonite shells, natural zeolites, clay, kaolinite and peat[6–12]. In addition, some other industrial waste by-products such as fly ash, waste slags and titanium oxide were also studied due to the economic benefits[13, 14]. Alumina has also found its way as an adsorbent in the water filtration industry thanks to its favorable characteristics like thermal stability, mechanical strength and adsorption capabilities. Because of its high porous structure and surface area, it is also used as a matrix to hold other materials for making of composites[15–17]. Furthermore, carbon nano tubes (CNTs) are well-known for the heavy metal removal capability[18–20]. This outstanding ability

IOP Conf. Series: Materials Science and Engineering **842** (2020) 012007 doi:10.1088/1757-899X/842/1/012007

of CNTs for the water purification applications stems from its inherent characteristics of highly porous structure, large specific surface area, strong affinity towards toxic pollutants, high thermo-chemical stability and excellent mechanical strength[21–24].

Thereby, preparation of porous ceramic membranes by using alumina matrix and CNT reinforcement seems to be a wise approach to achieve greater heavy metal removal efficiency. Most of the previous studies on alumina-CNT composites has been focused around dense composites primarily for the improvement in thermo-mechanical, electrical and tribological properties, however little work has been carried out towards the synthesis of porous alumina-CNT ceramic membranes[25–28]. The most commonly used technique to develop CNT based ceramic membranes is chemical vapor deposition (CVD), wherein CNTs are grown on ceramic substrates. Altalhi *et al.* developed CNT based membranes, wherein multiwalled CNTs were grown on alumina-polyamide substrate[29]. In another study Parham *et al.* developed porous alumina-CNT membranes by employing thermal pyrolysis technique[30].

Traditional ceramic powder processing method that involves formation of green pellets followed by sintering have been commonly employed to develop porous ceramic substrates having sufficient mechanical strength for ceramic membranes[31]. The final characteristics of the synthesized membranes are greatly influenced by the raw materials, processing methods, parameters and sintering techniques[32]. Not only this but the amount and type of binder used, initial compaction load and sintering cycle can also affect the physical and mechanical characteristics of the synthesized membrane[33, 34].

In the present work, powder metallurgy technique was utilized to develop alumina-MWCNT porous ceramic membranes for their intended use in water purification applications. Alumina-2wt.%MWCNT powder mixture was prepared using high energy ball milling technique followed by compaction using uniaxial pressing and further consolidation *via* solid state pressure less sintering. Effect of processing parameters namely compaction pressure and sintering temperature on the porosity, water permeability and mechanical strength of the synthesized membranes was investigated. Optimal processing parameters required to develop alumina-2wt.%MWCNT porous ceramic membrane having sufficient porosity with appropriate mechanical strength was identified.

2. Experimental procedure

α-alumina having a particle size of ~0.5 μm, procured from Buehler, Illinois, USA was used as the starting powder for the matrix. Multi-walled carbon nanotubes (MWCNTs) having a purity of 90% were procured from Times Nano, China, with an outer diameter of 10-20 nm and average length of 10-30 μm. Poly vinyl alcohol (PVA) was used as the binding agent during the compaction process. High energy ball milling (HEBM) was used for the preparation of alumina-MWCNTs powder mixture (2% loading of MWCNTs were added into 98% alumina matrix). The balls used in HEBM were made from zirconia with a weight ratio of one to twenty (for 1g of powder mixture 20 g of zirconia balls were used). Ethanol was used as a mixing media to attain proper mixing of alumina and MWCNTs. The mixture was prepared through HEBM at 3000 rpm for three hours. After three hours of mixing, the powder mixture was dried at 100 °C for about twenty-four hours in vacuum oven. Alumina-2%MWCNTs mixture was compacted into disc shaped pellets using a uniaxial powder compression machine with a holding time of about two minutes at three different pressures of 100 MPa, 250 MPa and 400 MPa. A stainless-steel die having a diameter of 27 mm was used for compaction. This resulted in green pellets of approximately 2.5-3 mm thickness and 27 mm diameter. The pellets were then taken out of the die with the help of a hydraulic press. The green pellets were then sintered at temperatures of 1200 °C, 1250 °C and 1300 °C so that the effect of variation of temperature could be studied on the synthesized membranes. Sintering was carried out in argon atmosphere to prevent oxidation of CNT at temperature values greater than 800 °C. During the sintering cycle, the heating rate was set at 5 °C/min and temperature was held constant at 500 °C for 4 hours followed by isothermal sintering for a period of 4 hrs at the respective sintering temperatures. The first batch of experiments included three green pellets compressed at 100 MPa followed by sintering at 1200 °C, 1250 °C and 1300 °C. Similarly, the next two batches were compressed at 250 MPa and 400 MPa and then sintered at the aforementioned temperatures.

IOP Conf. Series: Materials Science and Engineering **842** (2020) 012007 doi:10.1088/1757-899X/842/1/012007

Field emission scanning electron microscope (Lyra 3, Tescan, Czech Republic) was utilized to examine the microstructure of the initial powder mixture and the synthesized membranes. FESEM micrographs were acquired at an accelerating voltage of 20 kV in order to evaluate the effect of process parameters on the grain size of alumina matrix, size and distribution of the pores. Phase analysis was performed using X-ray diffraction (XRD), (Rigaku MiniFlex X-ray diffractometer, Japan). The diffractometer was operated at a wavelength, current and voltage of 0.15416 nm, 10 mA, and 30 kV respectively. Scan rate of 2°/min for 2 theta of 20° to 80° was used to analyze the crystal structure of the synthesized membranes for changes, if any. Water flux measurements were determined with the help of a custom designed test bench where the transmembrane pressure across the membrane housing was varied in the range of 1-4 bar. Water permeability of the membranes (l/m^2.hr) was calculated by $J = V/A \cdot t$, where A represents the cross-sectional area of the membrane, V represents the amount of water that flowed through the membrane in liters (l) and t is the time. The porosity of the membrane was measured as per the ASTM standards for porous ceramic materials (ASTM c373-14a). Suspended weights, dry weights and saturated weights were experimentally calculated using a Mettler Toledo kit with distilled water as the media. The porosity was calculated using the equations 1 and 2 and an average of five readings is presented in the results section.

$$Volume = saturated\ weight - suspended\ weight \tag{1}$$
$$Porosity = [(saturated\ weight - dry\ weight) / volume] \times 100 \tag{2}$$

The mechanical strength of the membrane was evaluated using diametrical compression test. This test was used to estimate the strength of porous ceramic membranes. In this test, the membrane was placed in vertical position between the two flat plates of compression testing machine and the tensile stresses developed within the membrane caused the failure of the membrane into two pieces. The crack traversing through the middle of the sample is an indication of successful completion of the test. The diametrical strength of membrane was calculated using the following equation.

$$\sigma = 2 \times (Applied\ Load) / (\pi \times Diameter \times Thickness) \tag{3}$$

Here σ is the amount of stress that the membrane can bear before permanent fracture. This is a useful value as it helps in determination of suitable membrane for practical applications with specific strength requirements.

3. Results and discussion

FESEM was performed for the starting alumina and raw MWCNT powders as well as the HEBM powder mixture of alumina/2wt.%MWCNT. Figures 1a and 1b show the FESEM micrographs of the as received alpha alumina and MWCNTs respectively. Figure 1c depicts FESEM micrographs of HEBM powder mixture. The comparison of the as received powders and HEBM powder mixture reveal that the alpha alumina particle size decreased from ~500 nm to ~270 nm due to HEBM. Furthermore, the clots of MWCNTs were observed to have opened up and fairly homogenous distribution of MWCNTs in the alumina matrix was evident. Figure 1d represent the XRD pattern of raw MWCNTs, starting alumina and HEBM powder mixture. The hump representative of MWCNTs was not observed in the XRD spectrum of HEBM powder mixture primarily due to the low weight % of CNTs in the mixture i.e. 2wt% (below the detection limit of XRD). Figure 1e presents the XRD patterns of HEBM powder mixture and the membranes synthesized at 1300 °C-100 MPa, 250 MPa and 400 MPa. In all the samples, alpha alumina peaks were seen to reside at the same 2-theta angles indicating that there wasn't any chemical reaction between the starting materials (alumina and MWCNTs) or that of the starting materials with the surrounding atmosphere.

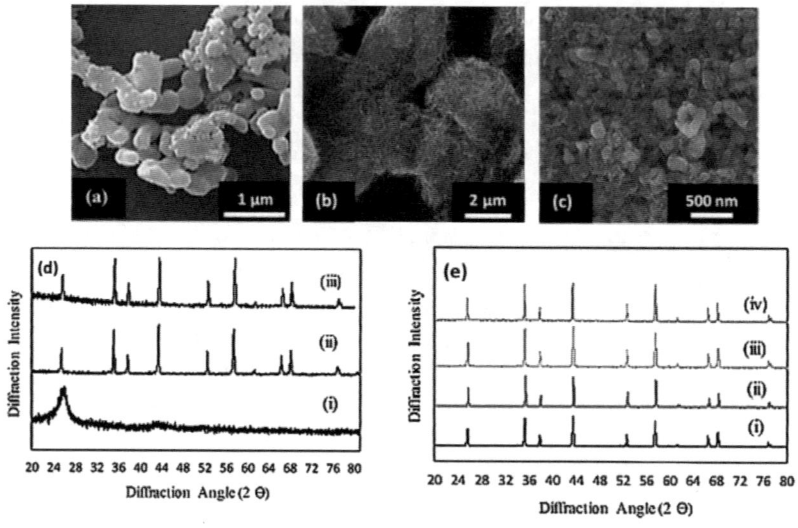

Figure 1. FESEM micrographs of (a) alpha alumina starting powder, (b) MWCNTs and (c) HEBM powder mixture of alpha alumina/2wt.% MWCNT. (d) XRD pattern of (i) MWCNTs, (ii) Alumina, (iii) HEBM powder mixture (98% alumina-2wt% CNTs) and (e) XRD pattern of (i) HEBM powder mixture (98% alumina-2wt% CNTs), (e) alumina/MWCNT disks sintered at 1300°C and compacted at (ii) 100MPa, (iii) 250 MPa, (iv) 400MPa.

Figures 2a and 2b represent the FESEM micrographs of the membranes sintered at 1200 °C. The membranes depict a porous and loosely fused morphology of alumina grains at low compaction pressure of 100 MPa that seems to improve further with the increase in the initial compaction load of 400 MPa (Figure 2b). Figures 2c and 2d show the micrographs of the membranes sintered at 1250 °C. The morphology of alumina particles appeared to be more compact with relatively lesser porosity as compared to the samples sintered at the lower temperature. The apparent porosity as well as the loosely held appearance of the of the matrix particles was observed to decrease with the increase in the initial compaction load from 100 MPa to 400 MPa. Figures 2e and 2f depict the micrographs of the samples sintered at 1300 °C. Grain growth of alumina particles at the expense of smaller grains is quite evident even at the lowest compaction pressure of 100 MPa. It is observed that with the increase in compaction pressure alumina particles fused well into each other and the apparent porosity is much less as compared to the samples prepared at the lower temperatures and compaction loads. As reported in the literature, the morphological evidence revealed that increasing the pressure caused the alumina particles to pack closely together which resulted in strong network development [35]. The greater contact area of grains allows the grains to fuse together more profoundly at higher compaction pressure as compared to lower compaction pressures.

IOP Conf. Series: Materials Science and Engineering **842** (2020) 012007 doi:10.1088/1757-899X/842/1/012007

Figure 2. FESEM micrographs of samples **(a)** and **(b)** sintered at 1200 °C and compacted at 100 MPa and 400 MPa respectively. **(c)** and **(d)** sintered at 1250 °C and compacted at 100 MPa 400 MPa respectively. **(e)** and **(f)** sintered at 1300 °C and compacted at 100MPa and 400 MPa respectively.

By increasing the temperature, porosity of the samples was observed to decrease. For 100 MPa samples, increasing the temperature from 1200 to 1250 to 1300 °C, the porosity was observed to decrease from 48.57% to 45.06% to 39.95% respectively. Similarly, the percentage porosity decreases with the increase in compaction pressure. Similar trend was observed for 250 MPa samples where the porosity decreased from 43.67% to 39.91% to 31.89% with the increase in sintering temperature. Further increase in compaction pressure to 400 MPa resulted in a relatively lesser reduction in porosity of 39.25% to 35.16% to 28.87%. Generally, for all the sintering temperatures, the reduction in porosity was more evident with the increase in compaction pressure from 100 MPa to 250 MPa followed by a relatively lesser decrease in porosity as the compaction pressure was increased from 250 MPa to 400 MPa.

Diametrical compression strength is the stress that a membrane can withstand under the application of a transmembrane pressure. The variation in diametrical strength of the synthesized membranes as the initial compaction pressure was increased. As it is already mentioned that the porosity decreased with the increase in initial compaction pressure. The growth of alumina grains as well as the fusion of grains was quite evident both at high sintering temperature as well as higher compaction load. The diametrical strength showed an increasing trend with the increase in the values of the process parameters. The measured strengths of the membranes prepared in the current study are well in agreement with the previously reported literature on diametrical compression strengths of CNT based membranes. The effect of process parameters (compaction loads and sintering temperatures) on the porosities and diametrical strengths of the synthesized membranes are summarized in table 1.

Table 1. Physical and mechanical properties of the synthesized membranes against respective process parameters.

S. No	Sample ID	Compactio n Pressure (MPa)	Sinterin g Temp. (°C)	Porosity (%)	Diametrical Strength (MPa)
1	130100	100	1300	37.95	5.35
2	130250	250	1300	31.89	7.59
3	130400	400	1300	28.87	11.07
4	125100	100	1250	45.06	2.87

IOP Conf. Series: Materials Science and Engineering **842** (2020) 012007 doi:10.1088/1757-899X/842/1/012007

5	125250	250	1250	39.91	5.11
6	125400	400	1250	35.16	7.35
7	120100	100	1200	48.57	0.95
8	120250	250	1200	43.67	2.28
9	120400	400	1200	39.25	4.52

Figures 3 a-c represent the water permiablility data of the synthesised membranes with refernce to the process parameters of sintering temperature and initial compaction loads. The water flux values of these membranes were determined for transmembrane pressure in the range of 1-4 bar. Increasing the transmembrane pressure result in an increase of water flux for all the membranes. As porosity and flow rates follow a direct relationship, therby with increase in the sintering tempereture (for same compaction load) the flow rates of the membranes decreased. With regards to the processing conditions, membranes sysnthesied using the compaction load of 100 MPa and sintering temperature of 1200 °C, high flow rates were recorded. The maximum flow rate of about 950 l/m^2.hr was achieved for the membrane processed at 100 MPa and1200 °C. The membranes synthesized in the current study exhibited higher flow rates as compared to the CNT based membranes developed through CVD method reported in literature[36].

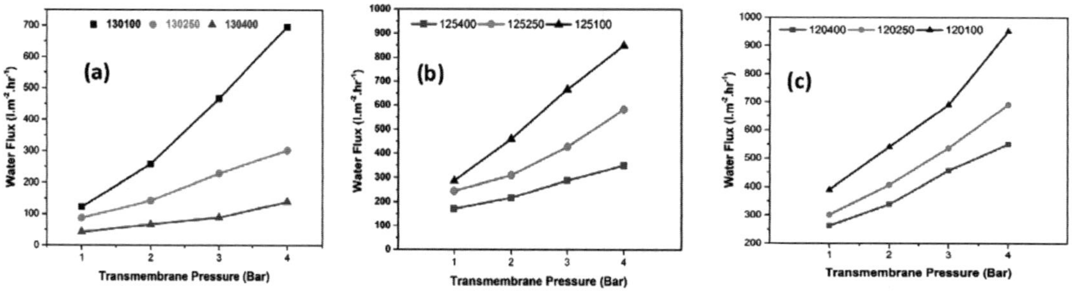

Figure 3. Effect of compaction pressure on water permeability for the membranes sintered at **(a)** 1300 °C, **(b)** 1250 °C and **(c)** 1200 °C.

4. Conclusion

In this work alumina/2wt.% MWCNT powder mixture was prepared using high energy ball milling process. The powder mixing process helped in achieving homogenous powder mixture of the alumina matrix and the MWCNTs. The process parameters, such as pressure (100 MPa, 250 MPa and 400 MPa) and sintering temperature (1200 °C, 1250 °C and 1300 °C) were found to greatly affect the physical and mechanical characteristics of the developed membranes. Water flow rates varied from 950 l/m^2·hr to 42 l/m^2·hr, porosity decreased from 48.57% to 28.87% and diametrical strength increased from 0.95 MPa to 11.06 MPa when compaction pressure was increased from 100 MPa to 400 MPa and sintering temperature increased from 1200 °C to 1300 °C. Membrane synthesized at 1200 °C, compacted at an initial load of 100 MPa, being able to withstand the transmembrane pressure of up to 4 bar and having the highest water permeability would now be subjected to assess the arsenic heavy metal removal capability.

5. References

[1] Mark J. Hammer, *Water and Wastewater Technology*, 5th ed. Prentice Hall, 2004.
[2] D. P. Li and Z. J. Wu, "Impact of chemical fertilizers application on soil ecological environment," *Chinese Journal of Applied Ecology*. 2008.
[3] A. Q. R Ullah, RN Malik, "Assessment of groundwater contamination in an industrial city, Sialkot, Pakistan," *African J. Environ. Sci. Technol.*, vol. 3, no. 12, 2009.

[4] R. A. Sial, M. F. Chaudhary, S. T. Abbas, M. I. Latif, and A. G. Khan, "Quality of effluents from Hattar Industrial Estate.," *J. Zhejiang Univ. Sci. B.*, 2006.
[5] K. G. Bhattacharyya and S. Sen Gupta, "Adsorption of a few heavy metals on natural and modified kaolinite and montmorillonite: A review," *Advances in Colloid and Interface Science*. 2008.
[6] X. Gu and L. J. Evans, "Surface complexation modelling of Cd(II), Cu(II), Ni(II), Pb(II) and Zn(II) adsorption onto kaolinite," *Geochim. Cosmochim. Acta*, 2008.
[7] S. A. Al-Jlil and F. D. Alsewailem, "Saudi Arabian clays for lead removal in wastewater," *Appl. Clay Sci.*, 2009.
[8] S. J. Köhler, P. Cubillas, J. D. Rodríguez-Blanco, C. Bauer, and M. Prieto, "Removal of cadmium from wastewaters by aragonite shells and the influence of other divalent cations," *Environ. Sci. Technol.*, 2007.
[9] D. Mohan and S. Chander, "Removal and recovery of metal ions from acid mine drainage using lignite-A low cost sorbent," *J. Hazard. Mater.*, 2006.
[10] A. Lu, S. Zhong, J. Chen, J. Shi, J. Tang, and X. Lu, "Removal of Cr(VI) and Cr(III) from aqueous solutions and industrial wastewaters by natural clino-pyrrhotite," *Environ. Sci. Technol.*, 2006.
[11] G. Sheng *et al.*, "Adsorption of Pb(II) on diatomite as affected via aqueous solution chemistry and temperature," *Colloids Surfaces A Physicochem. Eng. Asp.*, 2009.
[12] M. Betancur, P. R. Bonelli, J. A. Velásquez, and A. L. Cukierman, "Potentiality of lignin from the Kraft pulping process for removal of trace nickel from wastewater: Effect of demineralisation," *Bioresour. Technol.*, 2009.
[13] K. Ishizaki and M. Nanko, "A hot isostatic process for fabricating porous materials," *J. Porous Mater.*, 1995.
[14] V. K. Gupta, M. Gupta, and S. Sharma, "Process development for the removal of lead and chromium from aqueous solutions using red mud - An aluminium industry waste," *Water Res.*, 2001.
[15] I. Nettleship, "Applications of porous ceramics," *Key Eng. Mater.*, 1996.
[16] Y. C. Sharma, V. Srivastava, S. N. Upadhyay, and C. H. Weng, "Alumina nanoparticles for the removal of Ni(II) from aqueous solutions," *Ind. Eng. Chem. Res.*, 2008.
[17] M. J. Santos Yabe and E. de Oliveira, "Heavy metals removal in industrial effluents by sequential adsorbent treatment," *Adv. Environ. Res.*, 2003.
[18] K. Pillay, E. M. Cukrowska, and N. J. Coville, "Multi-walled carbon nanotubes as adsorbents for the removal of parts per billion levels of hexavalent chromium from aqueous solution," *J. Hazard. Mater.*, 2009.
[19] N. A. Kabbashi, M. A. Atieh, A. Al-Mamun, M. E. Mirghami, M. D. Z. Alam, and N. Yahya, "Kinetic adsorption of application of carbon nanotubes for Pb(II) removal from aqueous solution," *J. Environ. Sci.*, 2009.
[20] H. Wang, A. Zhou, F. Peng, H. Yu, and J. Yang, "Mechanism study on adsorption of acidified multiwalled carbon nanotubes to Pb(II)," *J. Colloid Interface Sci.*, 2007.
[21] N. M. Mubarak, J. N. Sahu, E. C. Abdullah, and N. S. Jayakumar, "Removal of heavy metals from wastewater using carbon nanotubes," *Sep. Purif. Rev.*, 2014.
[22] G. P. Rao, C. Lu, and F. Su, "Sorption of divalent metal ions from aqueous solution by carbon nanotubes: A review," *Sep. Purif. Technol.*, 2007.
[23] V. K. K. Upadhyayula, S. Deng, M. C. Mitchell, and G. B. Smith, "Application of carbon nanotube technology for removal of contaminants in drinking water: A review," *Science of the Total Environment*. 2009.
[24] X. Ren, C. Chen, M. Nagatsu, and X. Wang, "Carbon nanotubes as adsorbents in environmental pollution management: A review," *Chemical Engineering Journal*. 2011.
[25] J. W. An, D. H. You, and D. S. Lim, "Tribological properties of hot-pressed alumina-CNT composites," *Wear*, 2003.
[26] L. Kumari *et al.*, "Thermal properties of CNT-Alumina nanocomposites," *Compos. Sci. Technol.*, 2008.

[27] S. I. Cha, K. T. Kim, K. H. Lee, C. B. Mo, and S. H. Hong, "Strengthening and toughening of carbon nanotube reinforced alumina nanocomposite fabricated by molecular level mixing process," *Scr. Mater.*, 2005.

[28] F. Inam, A. Heaton, P. Brown, T. Peijs, and M. J. Reece, "Effects of dispersion surfactants on the properties of ceramic-carbon nanotube (CNT) nanocomposites," *Ceram. Int.*, 2014.

[29] T. Altalhi, M. Ginic-Markovic, N. Han, S. Clarke, and D. Losic, "Synthesis of carbon nanotube (CNT) composite membranes," *Membranes (Basel).*, 2010.

[30] H. Parham, A. Kennedy, and Y. Zhu, "Preparation of porous alumina-carbon nanotube composites via direct growth of carbon nanotubes," *Compos. Sci. Technol.*, 2011.

[31] P. Maarten Biesheuvel and H. Verweij, "Design of ceramic membrane supports: Permeability, tensile strength and stress," *J. Memb. Sci.*, 1999.

[32] M. Fukushima *et al.*, "Microstructural characterization of porous silicon carbide membrane support with and without alumina additive," *J. Am. Ceram. Soc.*, 2006.

[33] T. Mohammadi, A. Pak, Z. Nourian, and M. Taherkhani, "Experimental design in mullite microfilter preparation," *Desalination*, 2005.

[34] K. Shafiei, M. Kazemimoghaddam, T. Mohammadi, and S. Ghanbari Pakdehi, "An investigation on manufacturing of alumina microfiltration membranes," *Desalin. Water Treat.*, 2015.

[35] H. K. Shahzad, M. A. Hussein, F. Patel, N. Al-Aqeeli, M. A. Atieh, and T. Laoui, "Synthesis and characterization of alumina-CNT membrane for cadmium removal from aqueous solution," *Ceram. Int.*, 2018.

[36] M. Ahmadzadeh Tofighy and T. Mohammadi, "Nickel ions removal from water by two different morphologies of induced CNTs in mullite pore channels as adsorptive membrane," *Ceram. Int.*, 2015.

Acknowledgements

The authors would like to acknowledge the support provided by the Higher Education Commission (HEC) of Pakistan through the Startup Research Grant Program (SRGP) for funding this work through project No:21- 2415/SRGP/R&D/HEC/2019.The authors would also like to acknowledge the support provided by Dr. M. Shoaib for his help in the preparation of the ceramic membranes and Dr. Abbas Saeed Hakeem for his support in the characterization of the synthesized membranes.

IOP Conf. Series: Materials Science and Engineering **842** (2020) 012008 doi:10.1088/1757-899X/842/1/012008

Interesting Remarks on the Comparison of Organomodified Nanomontmorillonites in Fibre-Cement Nanohybrids

Styliani Papatzani [1, 2, a], Sotirios Grammatikos [3, b] and Kevin Paine [4, c]

[1]Hellenic Army Academy, Department of Mathematics and Engineering Sciences, Vari, Greece
[2]Hellenic Ministry of Culture and Sports, Directorate for the Restoration of Byzantine and Post-byzantine Monuments, Athens, Greece
[3]Group of Sustainable Composites, Department of Manufacturing and Civil Engineering, Norwegian University of Science and Technology, Gjøvik, Norway
[4]BRE Centre for Innovative Construction Materials, University of Bath, BA2 7AY, Bath, U.K

[a]spapatzani@sse.gr

Abstract. This paper presents the effects of the addition of (i) the organomodified nanomontmorillonite (nMt) dispersion, nC2, produced for the scope of the FP7 project name FIBCEM and of (ii) the powder, undispersed organomodified industrially produced nMt, nC4 on quaternary nMt-fibre reinforced binders. The reference binder consisted of 60% Portland cement (PC), 20% limestone (LS) and 20% fly ash (FA). Superplasticizer was added at 2%, polyvinyl alcohol fibres at 3% and nMt was added at 1%. Flexural strength test were carried out at day 7, 28, 56 and 90 showed marginal enhancement offered by the nMt dispersion, which was attributed to limited, but existing additional pozzolanic reactivity of the particles of nMt. If fact, following methodology presented in published papers on other formulations, thermal gravimetric and X-ray diffraction analysis showed limited consumption of $Ca(OH)_2$ towards production of additional C–S–H, which is in line with the marginal increase of flexural strength for the same age of testing.

1. Introduction

Through a series of publications the effect of nanoparticles on composite cement pastes and binders was investigated [1, 2] in light of the necessity to lower the CO_2 footprint of cementitious composites. The greatest part of these studies covered ternary, quaternary or quinary pastes, which did not include fibres or superplasticizers, whereas two articles have been published on the use of nanomontmorillonites (nMt) on a non-pozzolanic reference formulation (containing Portland cement and limestone) of fibre-cement nanohybrids, which contained Portland cement and limestone [3, 4]. It was found that of the organomodified nMt's the best performing was the one aqueous dispersed in water with the help of an alkyl aryl sulfonate surfactant. The industrial powder nMt, nC4, did not offer any strength, chemical, or microstructural enhancements [4] and the inorganic nMt was presented as a promising nanoparticle [3].

In this work, research is being extended to quaternary formulations. Fly ash is added to the abovementioned cementitious matrix (now containing Portland cement, fly ash and limestone) in an

IOP Conf. Series: Materials Science and Engineering **842** (2020) 012008 doi:10.1088/1757-899X/842/1/012008

effort to lower the environmental impact of the reference paste. Then, nMt is added at 1% by total mass of binder, which has been found to be the optimum amount in previous research on cement pastes [5] or on fibre-cement nanohybrids [4]. The focus on this study has been to witness any enhancements of flexural strength by the addition of two different organomodified nMt's and to interpret the strength results with the help of thermogravimetrical and crystallographic findings.
To the best knowledge of the authors, this is the first time that an industrial organomodified powder nMt is compared against an organomodified aqueous dispersion nMt in low PC clinker quaternary fibre-cement nanohybrids.

2. Materials, mix design and methods

2.1. Materials
The materials used were:
- Portland limestone cement CEMII/A-L42.5, with a limestone content of 14%, conforming to EN 197-1. The supplier gave the following clinker composition: 70% C_3S, 4% C_2S, 9% C_3A, 12% C_4AF.
- Limestone (LS) (additional), conforming to EN 197-1.
- Fly ash (FA), conforming to EN 450. The oxide composition provided by the material data sheet was: 53.5% SiO_2, 34.3% Al_2O_3, 3,6% Fe_2O_3, 4.4% CaO.
- Organomodified nMt, nC2 dispersed in water with the help of an alkyl aryl sulfonate surfactant, containing about 15% by mass of nMt particles.
- Organomodified nMt; nC4 in powder form containing about 15% by mass of nMt particles.

2.2. Mix design
In the present paper a quaternary formulation comprising a pozzolanic reference paste containing 60%PC, 20%LS, 20%FA, 3% PVA fibres and 2% superplasticizer, denoted as F.PC60LS20FA20PVA3SP2 (F is for flexure) is enhanced with nMt. Water content in nMt dispersions accounted for 85%, which entered as part of the water content in the cement formulations to obtain water to binder ratio of 0.3 (or in other words water-to-solids ratio). Therefore, the general formula of the matrix of the resulting ternary nanomodified cement formulations was:

$$PC\ (60) + LS\ (20\text{-}x) + FA\ (20) + PVA\ (3) + SP\ (2) + x nMt \qquad (1)$$

Where x = % of nMt solids.
The mixing procedure was standardized by the authors. Pastes were cast in prismatic moulds, producing slabs of the following dimensions: 10 mm depth, 40 mm breadth and 120 mm length (figure 1) and compacted at a shaking table. They were subjected to a 24-hour air-curing and subsequently demoulded and cured in distilled water thereafter at 20±2 °C until the day of mechanical testing.

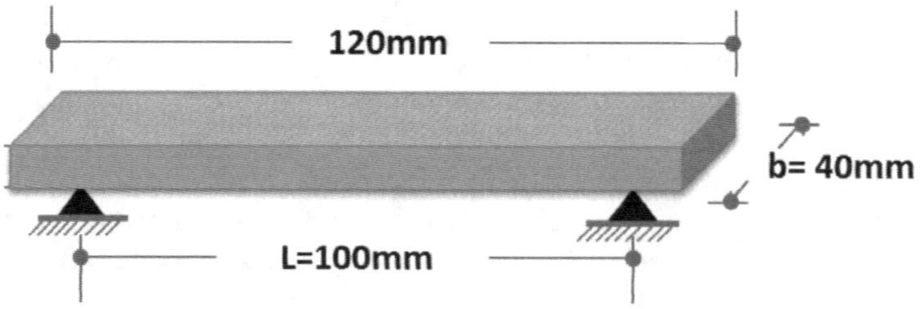

Figure 1. Dimensions of specimens.

2.3. Testing methodology

Flexural (three-point bending) strength tests were carried out in accordance with BS EN 12467. Mean strength values of three specimens were calculated, as well as standard deviation at day 7, 28, 56 and 90.

For the TGA/dTG and XRD paste characterization, arrest of hydration was performed following the oven drying technique, as described by Calabria-Holley et al. [6].

Thermogravimetric analysis (TGA) was carried out using a Setaram TGA92 instrument. Approximately 20 mg of nC4 were placed in an alumina crucible and heated at a rate of 10°C/min from 20°C to 1000°C under 100 mL/min flow of inert nitrogen gas. The differential thermogravimetric curve (dTG) was derived by the TG curve. The first derivative curve was produced for the various samples tested and was used for comparisons instead of the mass loss curve, as it yields sharp distinctive peaks.

XRD measurements were performed using a D8 ADVANCE X-ray diffractometer with CuKα radiation. Spectra were obtained in the range $4° < 2\theta < 20°$ at an angular step-size of 0.016° 2θ. The state and extent of dispersion - exfoliation of the NMt can be examined by XRD and TEM analysis with monitoring changes in basal spacing [7].

3. Results and discussion

3.1. Mechanical testing

Figure 2 displays the flexural strength of the different specimens. nC4 provided marginal strength improvement at early ages, which, however, showed a reduction in strength at later ages (90 days). What is also worth noting is that the standard deviation for the reference paste ranged from 1 to 1.2 MPa, however for nC2 modified samples, it reached approximately 1.2 MPa, whereas for nC4 it was much lower and equal to about 0.8 MPa. It can, therefore, be postulated that for cementitious nanohybrids it is best to disperse the nMt in water.

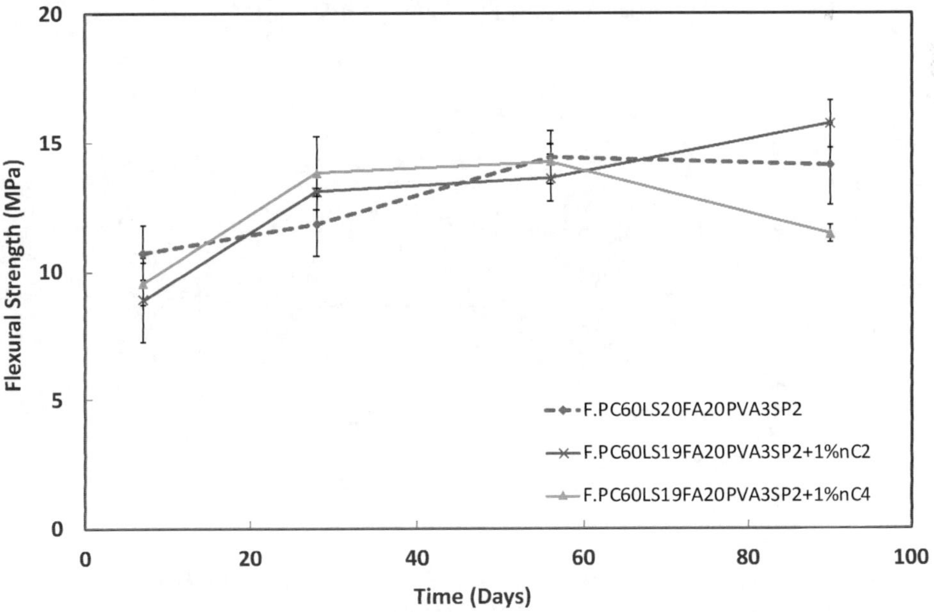

Figure 2. Flexural strength of 1% nC2 and 1% nC4 fibre-cement nanohybrids based on F.PC60LS20FA20PVA3SP2.

3.2. Thermal gravimetric and X-ray diffraction analysis

As shown in figure 3, at day 28, no additional C–S–H was produced by nC4, whereas the addition of nC2 produced limited quantities of additional ettringite and C–S–H [8]. According to other research results it is expected that as time proceeds, nC2 will consume more $Ca(OH)_2$ towards the production of additional C–S–H [4, 9, 10] .

In addition to this, X-ray diffraction analysis was also carried out at day 28 to confirm the consumption of $Ca(OH)_2$ and monitor the amount of $CaCO_3$ present in the formulations. Indeed, as shown in figure 4, the $Ca(OH)_2$ present in the reference paste was consumed with the addition of nMt. Again there is a high correlation between the $Ca(OH)_2$ consumed and the flexural strength [5, 8]. nC4 at day 28 seems to have delivered the highest flexural strength and has also consumed the greatest amount of $Ca(OH)_2$ although from the analysis between 100 – 180°C no additional C–S–H was produced.

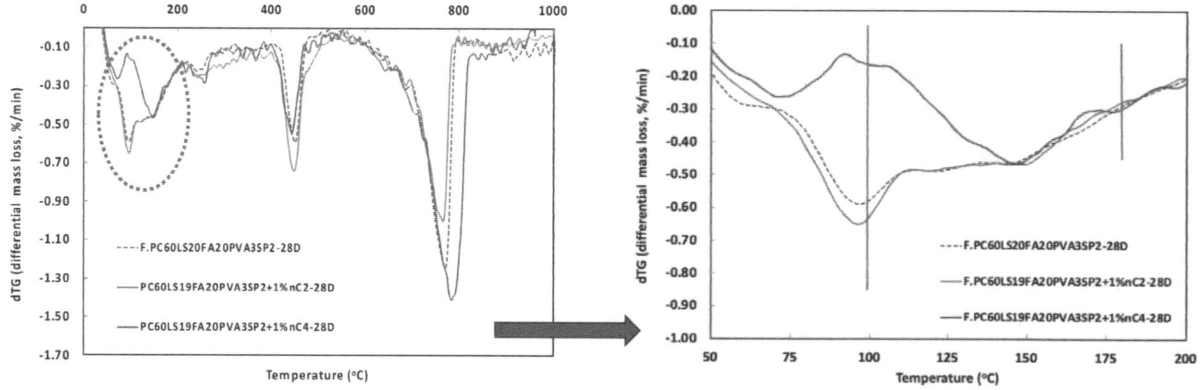

Figure 3. Differential mass loss between 100-200°C, at day 28 of fibre-cement nanohybrids based on F.PC60LS20FA20PVA3SP2.

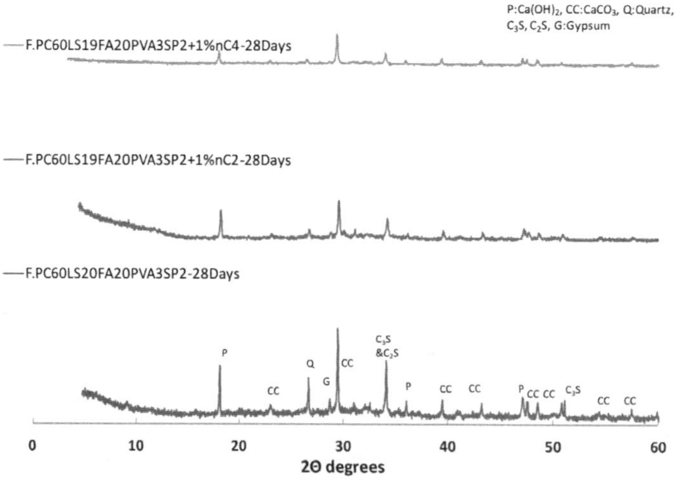

Figure 4. XRD pattern at day 28 of fibre-cement nanohybrids based on F.PC60LS20FA20PVA3SP2.

4. Conclusions

The effects of the addition of two completely differently produced organomodified nanomontmorillonites was assessed in terms of flexural strength. Results were correlated with the consumption of $Ca(OH)_2$ although limited production of additional C–S–H was attested. The overall consensus is that nMt performs better if dispersed in water, when cementitious nanocomposites are produced.

5. References

[1] S. Papatzani, Effect of nanosilica and montmorillonite nanoclay particles on cement hydration and microstructure, Mater. Sci. Technol. 32 doi:10.1179/1743284715Y.0000000067, 2016.

[2] J. Calabria-Holley, S. Papatzani, B. Naden, J. Mitchels, K. Paine, Tailored montmorillonite nanoparticles and their behaviour in the alkaline cement environment, Appl. Clay Sci. 143 (2017) 67–75. doi:http://dx.doi.org/10.1016/j.clay.2017.03.005, (2016) pp. 138–153.

[3] S. Papatzani, K. Paine, Dispersed Inorganic or Organomodified Montmorillonite Clay Nanoparticles for Blended Portland Cement Pastes: Effects on Microstructure and Strength, in: K. Sobolev, S.P. Shah (Eds.), Nanotechnol. Constr. Proc. NICOM5, Springer International Publishing, doi:10.1007/978-3-319-17088-6_16, (2015) pp. 131–139.

[4] S. Papatzani, K. Paine, From Nanostructural Characterization of Nanoparticles to Performance Assessment of Low Clinker Fiber–Cement Nanohybrids, Appl. Sci. 9. doi:10.3390/app9091938, 2019.

[5] S. Papatzani, K. Paine, Lowering cement clinker: A thorough, performance based study on the use of nanoparticles of SiO_2 or montmorillonite in Portland limestone nanocomposites, Eur. Phys. J. Plus. 133. doi:10.1140/epjp/i2018-12305-6, 2018.

[6] J. Calabria-Holley, K. Paine, S. Papatzani, Effects of nanosilica on the calcium silicate hydrates in Portland cement–fly ash systems, Adv. Cem. Res. 27. doi:10.1680/adcr.13.00098, (2015) pp. 187–200.

[7] S. Papatzani, K. Paine, Inorganic and organomodified nano-montmorillonite dispersions for use as supplementary cementitious materials - A novel theory based on nanostructural studies, Nanocomposites, doi:10.1080/20550324.2017.1315210, (2017) pp. 2–19.

[8] S. Papatzani, K. Paine, Polycarboxylate/nanosilica-modified quaternary cement formulations – enhancements and limitations, Adv. Cem. Res. doi:10.1680/jadcr.17.00111, (2018) pp. 256–269.

[9] S. Papatzani, K. Paine, Optimization of Low-Carbon Footprint Quaternary and Quinary (37 % Fly Ash) Cementitious Nanocomposites with Polycarboxylate or Aqueous Nanosilica Particles, Adv. Mater. Sci. Eng. doi:10.1155/2019/5931306, 2019.

[10] S. Papatzani, E.G. Badogiannis, K. Paine, The pozzolanic properties of inorganic and organomodified nano-montmorillonite dispersions, Constr. Build. Mater. V. 167, doi:10.1016/j.conbuildmat.2018.01.123, (2018) pp. 299–316.

Acknowledgments

This research was funded by the EU-FP7 FIBCEM project, Grant number: 262954. All partners of the FIBCEM project are thanked for their input and for the supply of materials. The authors would also like to acknowledge the Department of Chemical Engineering at the University of Bath for the use of the TG analyzer.

IOP Conf. Series: Materials Science and Engineering **842** (2020) 012009 doi:10.1088/1757-899X/842/1/012009

Thermal Insulation Enhancement of Metakaolin-Based Geopolymer Concrete Using Waste Clay Brick

Mahmood F. Ahmed[1,*], Wasan I. Khalil[2], Qias J. Frayyeh[3]

[1, 2, 3] Civil Engineering Department, University of Technology, Baghdad, Iraq

*E-mail: mahmodmfa81@gmail.com

Abstract. Geopolymer concrete is one of the sustainable building materials that has a low energy production and a minimum carbon footprint. In this study, the effect of utilization waste clay brick on the thermal conductivity (TC) of Metakaolin geopolymer concrete (MK-GPC) was investigated. The waste of clay brick was used in two series of mixes, first one included clay brick waste powder (BP) as partial replacement of Metakaolin (MK) at weight dosage 10%, 15%, and 20%, while the second group of mixes contained crushed clay brick waste aggregate (CBA) at replacement level of 10%, 20% and 30% by volume of natural coarse aggregate. In addition to (TC), all the specimens were tested to determine the ultrasonic pulse velocity (UPV), and voids content at 7 and 28 days. The results of the samples with 10% and 15% BP have shown higher UPV and less voids content, while the incorporation of CBA aggregate at 20% and 30% caused reduction in UPV and increased the voids in MK-GPC mixture. Furthermore, the addition of the waste clay brick in both forms (powder and aggregate) has improved the results of thermal insulation. Conclusively, reusing of waste clay brick as ingredients in MK-GPC will provide a super-sustainable geopolymer concrete with advanced thermal insulation properties.

1. Introduction

The growing of awareness about the environmental problems and energy consumption has made the sustainability in the sector of building and construction as an importance issue worldwide. The increasing demand on building materials, especially Ordinary Portland Cement (OPC) concrete and clay brick units has caused negative impacts related to energy consumption, environmental pollution and depletion of natural recourses. The global demand of OPC was over 3 billion tones in 2012, and this number will increase to 4.4 billion tones by the year 2050 [1]. In the production process, each ton of OPC releases one-ton of carbon dioxide (CO_2). This is directly contribute to the emission of greenhouse gases approximately 5-8%, and this amount could jumped up to 15% by 2020 [2]. Thus, many endeavor have been attentive on developing an eco-friendly innovative alternative to ordinary Portland cement commonly known as “Geopolymers” which firstly coined by Davidovits in 1978. Geopolymer concrete can be manufactured by chemical reaction of alkaline solution liquid; usually consist of sodium silicate (Na_2SiO_3) and sodium hydroxide (NaOH), with rich aluminosilicate-based powder materials such as fly ash, slag, red mud, and rice husk ash; or metakaolin (MK). On the other hand, the clay brick has considered as a second level of building materials after concrete. Huge amounts of waste clay brick are generating during the manufacturing, construction and demolition activities, and this type of waste has treated as C&D waste. Most of waste brick are left as landfill or illegally dumped, and become a serious environmental problem [3]. Another environmental drawback of construction field has come from high energy consumption for cooling and heating (thermal comfort) of the buildings, which is depending on thermal conductivity value of construction materials

[4]. Kim and Lee [5] have reported great (TC) value of 0.46 W/m.K and compressive strength 16.2 MPa at 28 days for fine ground bottom ash geopolymer samples. Meanwhile, Jaya et al. [6] have depended on optimum $Na_2SiO_3/NaOH$ ratio to produce metakaolin geopolymers with 32 MPa compressive strength and (TC) up to 0.82 W/m.K. Hence, the novelty of this research is toward minimizing the waste clay brick pollution throw reusing and recycling it as ingredients in synthesis more thermal insulation geopolymer concrete. The objectives of this study are to investigate the effects of different contents of BP and CBA aggregate on the (TC), (UPV) and voids content of MK-GPC specimens, and show the feasibility of production superior green-thermal insulation geopolymer concrete on the way to developing the sustainable construction industry.

2. Experimental Procedure

2.1. Materials

Iraqi kaolin clay was burning at 700 °C for 2 hours to produce metakaolin that used in this study as source binder. The alkaline activator was obtained from mixing 14M solution of (NaOH) with (Na_2SiO_3) at ratio of 1:2 for all mixes. The (BP) and (CBA) have produced by crushing the residual of non-used units of hollow clay brick into small pieces using crushing machine. The crushed particles with size 5-14 mm have collected separately and grading on electrical sieves shaker to be used as coarse aggregate, while the finer materials were re-grinded by cyclone machine to produce BP as shown in Figure 1 and 2 respectively. The chemical composition and physical properties for MK and BP are illustrated in Table 1. For all mixes, natural coarse aggregate with 14 mm maximum size; and natural sand with 3.3 fineness modulus were used. Also, extra tap water and superplasticizer (Conplast 2000) were added to improve the workability and mixing process.

Figure 1. Clay brick waste aggregate (CBA).

Figure 2. Clay brick waste powder (BP).

Table 1. Characteristics of Metakaolin (MK) and brick powder (BP).

Oxide %	SiO_2	Al_2O_3	Fe_2O_3	CaO	MgO	SO_3	Na_2O	K_2O	L.O.I
MK	54.20	39.0	0.92	1.37	0.15	0.45	0.22	0.27	0.71
BP	56.82	11.36	2.36	20.20	3.02	0.83	0.86	0.86	1.19

Physical Properties	**MK**	**BP**
Specific gravity	2.64	2.84
Specific surface area (m^2/gm.)	14.3	0.462
7 days pozzolanic activity index, (%)	113.3	89.2

2.2. Mix proportions and specimens preparation

The mix design for reference mixture was based on several trial mixes to produce MK-geopolymer concrete with compressive strength reach up to 40 MPa at age of 7 days. For all mixes, the alkaline solution to binder ratio was fixed at 0.65, while the dosage of superplasticizer and extra water were 2% and 10% by weight of metakaolin respectively. As summarized in Table 2, two series of mixes were prepared to understand the effect of waste clay brick on reference mix, the first one included replacing the weight of MK with BP at dosage of 10%, 15% and 20%, while the second series incorporated the CWCB aggregate as partial volumetric replacement of natural coarse aggregate at levels of 10%, 20%

and 30%. All specimens were cured by sealing it with a thick nylon bags, then placed in an electrical oven at 60 °C for 4-5 hours, and exposed to sun light in summer season at 36-48 °C until the test time.

Table 2. Mix proportions of all MK-GPC mixtures.

Mixes		Materials (kg/m^3)					Compressive Strength at 7 Day (MPa)
		MK	BP	CBA	Coarse Agg.	Fine Agg.	
Ref. MK-GPC		415	0	0	1240	475	40.39
Series I	10% BP	373.5	41.5	0	1240	475	22.09
	15% BP	352.75	62.25	0	1240	475	29.26
	20% BP	332	83	0	1240	475	25.36
Series II	10% CBA	415	0	73.08	1116	475	39.0
	20% CBA	415	0	146.16	992	475	26.83
	30% CBA	415	0	219.24	868	475	20.0

2.3. Test methods

The details of experimental tests required to recognizing the properties of all MK-GPC mixes are reported in Table 3. For each tests, three specimens were used to calculate the average of final result.

Table 3. Details of testing methods.

No.	Type of test	Geometry of sample	The Standards
1	Ultra Pulse Velocity (UPV)	Prism of 100×100×400 mm	ASTM C597-2015 [7]
2	Voids content	Fractured part of the tested prisms	ASTM C642-2013 [8]
3	Thermal Conductivity	Prism of 75×75×300 mm	ASTM C 1113-2013 [9]

3. Results and discussion

3.1. Ultrasonic pulse velocity

The the (UPV) test was conducted at 7 and 28 days to evaluate the matrix uniformity of MK-GPC mixes that containing different forms and percentages of waste clay brick. The results of UPV test for series I and series II are shown in Figure 3 and 4 respectively. With continuous of geopolymerization process the microstructure would enhance, thus the UPV values for all mixes have increasing with time

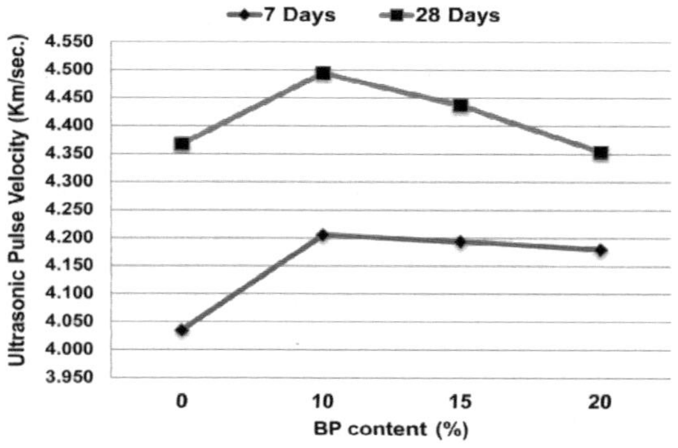

Figure 3. UPV of MK-GPC with BP.

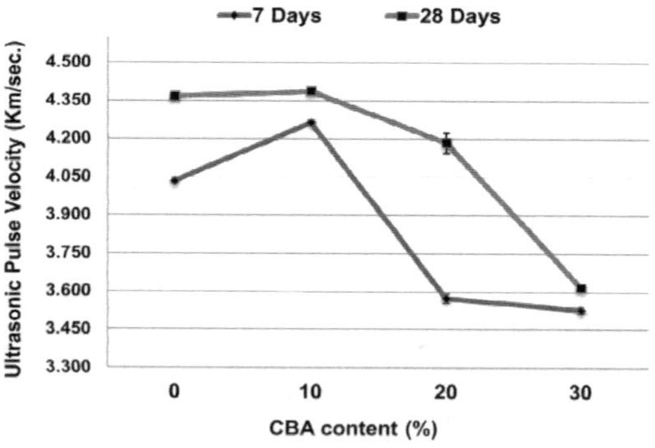

Figure 4. UPV of MK-GPC with CBA.

As appeared from Figure 3, the pulse velocity of 10% and 15% BP mixes were slightly higher than reference mix. This is because of the filling effect of waste brick powder and facilitated pore refinement of the geopolymer matrix. On the other hand, Figure 4 shows that the mix of 10% CBA has a smooth increase in UPV result, on contrary of that, the incorporate of 20% and 30% CBA declined the pulse velocity of reference mix up to 4.25% and 17.24% respectively. This behaviour attribute to the porous structure of crushed clay brick aggregate and weakness of inartificial transition zone.

3.2. Voids content

Voids content data of MK-GPC containing BP and CBA are shown in Figure 5 and 6 respectively. It can be seen that voids content for all mixes are decreased with time due to continuous geopolymerization activity and development of the microstructure of binder paste. The voids content in mixes with 10%, 15% and 20% BP are less than that of reference mix by 21.19%, 30.64% and 18.35% respectively. These outcomes are because of filling action of BP and the formation of compacted binder in the geopolymer matrix [10], that is comply with the UPV findings. In same trend, the incorporation of 10% CBA has minimized the voids in MK-GPC by 5.87%, while the dosages of 20% and 30% of CBA has increased the voids content of control mix up to 11.56% and 27.98% respectively. The porous and cellular structure of CBA has produced more voids in MK-GPC matrix.

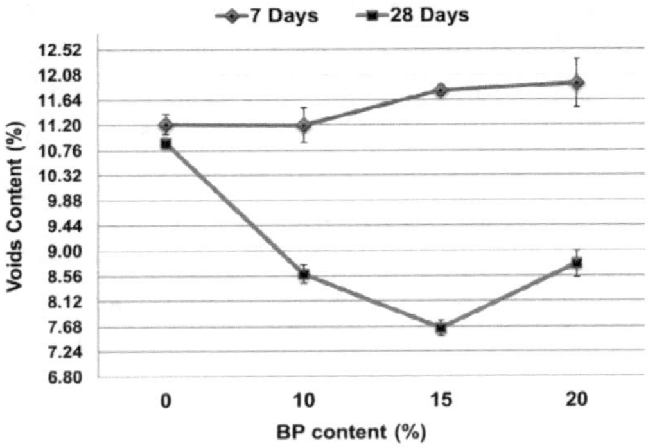

Figure 5. Voids content of MK-GPC with WBP.

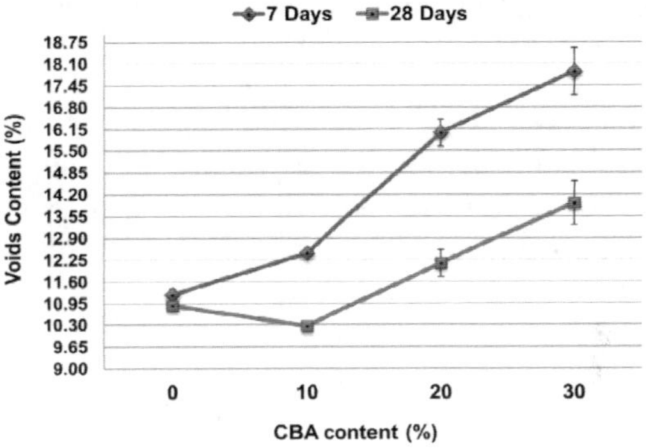

Figure 6. Voids content of MK-GPC with CBA.

3.3. Thermal conductivity

Figure 7 is given the thermal conductivity values for all MK-GPC mixes at age of 28 days. Thermal conductivity of reference mix (0% waste clay brick) was 1.5295 W/m.K. This value decreased by 42.90%, 45.00%, and 46.27% for mixes with 10%, 15% and 20% BP, respectively.

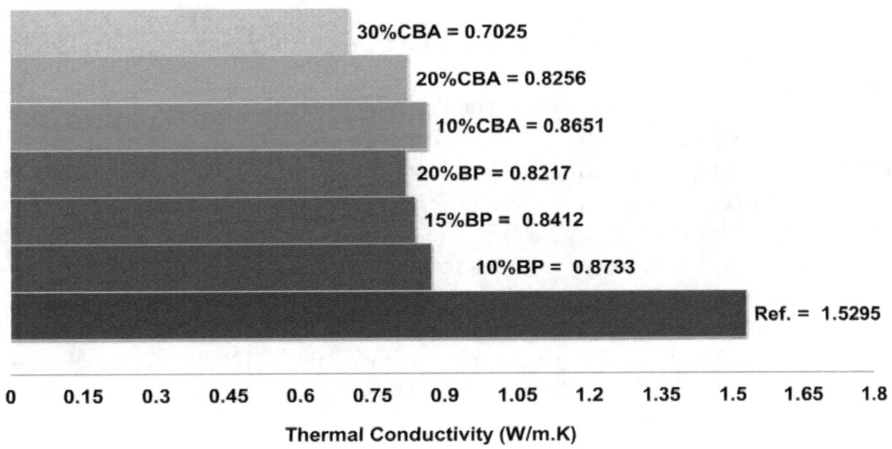

Figure 7. Results of thermal conductivity for all MK-GPC mixes.

In spite of the voids content results which indicate that BP-mixes have less voids and pores, but the incorporation of BP in MK-GPC has minimized the thermal conductivity. This can be attributed to the nature and insulation characteristic of the clay brick powder compared with metakaolin. Furthermore, with replacement natural coarse aggregate by CBA, the thermal conductivity of control mix has intensively declined up to 43.44%, 46.02% and 54.07%, for mixes with 10%, 20% and 30% CBA, respectively. In compared with natural coarse aggregate, the CBA has lower coefficient of thermal conductivity [11], more porous and cellular structure. This will produced more pores and voids in the concrete matrix, leading to lowering (TC) of reference mix and enhanced the thermal insulation. Although the MK-GPC was a normal weight concrete, but the obtained (TC) for all mixes were in close to lightweight geopolymer concrete which range between 0.620 to 0.957 W/m.K [12].

4. Conclusions

In this study, the waste clay brick was utilized as a part of source binder and coarse aggregate to produce sustainable geopolymer concrete. The main findings of using waste clay brick powder in MK-GPC at optimum ratio 15wt.%, are improvement of the pulse velocity and minimization the voids content in the geopolymer paste. On contrary of that, the using of waste brick aggregate has a negative effect on the MK-geopolymer matrix, and that led to reduce the ultrasonic velocity and increase the voids content in the paste. Moreover, the results of thermal conductivity shows a great enhancement when the waste clay brick has added to the MK-GPC mix. The replacement of MK at 20wt. % of BP has provided the best level in thermal insulation results, whereas the dosage of 30 vol. % of CBA has the optimum minimizing of the thermal conductivity to reach 0.7025 W/m.K. The key findings of this investigation confirm the importance of using the waste of clay brick as a main material to synthesis more sustainable geopolymer concrete with superior performance in thermal insulation. This will reduce energy consumption during manufacturing and life cycle of the buildings, as well as, minimize the environmental pollution, CO_2 emission and exhaustion of raw materials. Further researches on micro-scanning structure, durability performance and time dependent deformation for MK-GPC with waste clay brick are suggested to be investigated comprehensively.

5. References

[1] Riccardo M, Jennifer J R and Andrea H 2018 *Journal of Cleaner Production.* **186** 933-942
[2] Guillaume H and Claudiane OP 2016 *RILEM Technical Letters.* **1** 17-23
[3] Chee LW, Kim H M, Soon P Y, Johnson U A and Tung CL2018 *Journal of Cleaner Production.* **195** 226-239
[4] Iman A, Payam S, Zahiruddin F BA and Norhayati BM 2018 *Journal of Building Engineering.* **20** 81-93
[5] Hyo Kim and Ji-Young Lee 2017 World of Coal Ash (WOCA) conference in Lexington, KY- 9-11 May, 2017
[6] Nur A J, Liew Y-M, Mohd M A and Heah C-Y 2018 *Solid State Phenomena.* **280** 487-493
[7] ASTMC597-2015 Standard test method for Pulse Velocity through concrete, American Society for Testing and Materials; 2015.
[8] ASTM C642–2013 Standard test method for density, absorption, and voids in hardened concrete, American Society for Testing and Materials, 2013
[9] ASTM C1113–2009 (Reapproved 2013) Standard test method for thermal conductivity of refractories by hot wire (platinum resistance thermometer technique), American Society for Testing and Materials, 2013
[10] Ivana K, Ivanka N, and Dubravka B2008 Technical *Gazette.* **15** 35-40
[11] BrînduşS, Vyšvařil M, Bayer P, Keppert M and RovnaníkovàlP 2018 International Conference Building Materials, Product and Technologies, Blansko-Češkovice, Czech Republic *Materials Science and Engineering.* **379**1-8
[12] Wasan I K, Waleed A A and Ibtesam F N 2018 1st International Scientific Conference of Engineering Sciences-3rd Scientific Conference of Engineering Science (ISCES). 10-11 Jan. Diyala- Iraq 175-180

IOP Conf. Series: Materials Science and Engineering **842** (2020) 012010 doi:10.1088/1757-899X/842/1/012010

Wear Characteristics by Ti Coating of Al 7075

Jihye An[1], Sunghoon Im[2], Yunyoong Yoo[3], Jungpil Noh[4], Sunchul Huh[4*]

[1]Dept. of Energy and Mech. Eng. Graduate school of GNU, Tongyeong-si, Korea
[2]Dept, of Computer Applied Machinery Changwon Campus of Korea Polytechnic, Changwon-si, Korea
[3]KP Aero Industries Co., LTD Gimhae-si, Korea
[4]Dept. of Energy and Mech. Eng. Of GNU, Tongyeong-si, Korea

[*]schuh@gnu.ac.kr

Abstract. Al 7075 alloys are widely used in many mechanical parts because of their low density, light weight, and corrosion resistance. However, when mechanical parts are used in extreme environments such as high speed, high temperature, and high pressure, wear occurs more than allowable. To solve this problem, research and development of surface thin film deposition technology are increasing. Therefore, in this study, the wear characteristics were studied by depositing Ti films with excellent mechanical properties on Al 7075 according to the surface roughness by the deposition time. Results showed that 60' sputtered specimens after polishing with alumina had the best wear resistance. In addition, the surface roughness of the base material and the deposition time of the thin film affected the wear resistance.

1. Introduction

Recently, extensive research and development have been focused on various materials such as aluminum, magnesium, and functional plastics to achieve weight reduction in automotive and machinery industries [1]. Among them, aluminum is a representative material applied for the purpose of weight reduction of products because it has about 1/3 of iron specific gravity. In particular, Al 7075 is widely used in various mechanical parts because of its excellent strength, corrosion resistance, and thermal conductivity [2].

However, when mechanical parts are used in extreme environments such as high speed, high temperature, and high pressure, wear occurs more than allowable [3]. To solve this problem, research and development on surface thin film deposition techniques have been increased [4, 5]. Surface thin film deposition technology is a method for improving mechanical properties by depositing a material with excellent wear characteristics on the surface of materials [6, 7]. Therefore, mechanical properties should be considered when depositing a thin film on a material. In addition, the effect of surface roughness of the base material and the deposition time of the thin film on wear characteristics need to be studied.

Thus, the objective of this study was to investigate wear characteristics after depositing Ti with excellent wear resistance on Al 7075 with various surface roughness values through DC magnetron sputtering.

2. Material and methods

In this study, a total of 13 samples of Al 7075 with diameter of 32 mm and height of 10 mm were prepared as base materials. In order to know the wear resistance according to the surface roughness of the base material, the surface was polished with # 400, # 800, # 1200, and #Alumina, respectively. The roughness was then measured.

After measuring the surface roughness, each specimen was deposited with a Ti thin film by time. Table 1 shows conditions used to deposit a thin film. All conditions except deposition time were the same. Twelve specimens were prepared by varying the deposition time according to roughness. In addition, each specimen was compared based on the base material (13) for which thin film was not deposited. Conditions used for each specimen are summarized in table 2.

For each specimen, wear characteristics were determined using a TRIBOSS PD-102 abrasion testing machine. Experimental method was carried out with a ball on disk method using a zirconium with ball diameter of 12.7 mm. The rotational speed of the ball was 60 RPM. The experimental time was 30 minutes and the load was set to be 0.2 kg. The wear test can be used to calculate the friction coefficient and the abrasion loss. A scanning electron microscope (SEM) was used to observe the resulting wear track.

3. Result and discussion

3.1. Surface roughness measurement

Table 3 shows surface roughness value of the base material for each specimen. Average value was obtained from 15 measurements for each specimen. The higher the particle size, the lower the roughness value.

3.2. Friction coefficient analysis

Figure 1 is a graph showing the coefficient of friction according to the roughness of the Ti thin film deposition time. In the graphs (a), (b), and (c), the friction coefficient decreased as the deposition time increased, but the friction coefficient of the 90' sputtered specimen was higher in the graph. It is thought that such a phenomenon occurred because the thin film was excessively deposited and the bond ability with the base material was reduced.

Table 1. Sputtering conditions.

Film Material	Deposition Time	Base Vacuum [torr]	Working Vacuum [torr]	Plasma Factor [W]	Temperature
Ti	Each Conditions	5.0×10^{-6}	2.0×10^{3}	200	RT

Table 2. Deposition time according to surface roughness.

		Surface roughness			
		#400	#800	#1200	#Alumina
Deposition Time	30'	1	2	3	4
	60'	5	6	7	8
	90'	9	10	11	12

Table 3. Average surface roughness value [Ra].

#400	#800	#1200	#Alumina
0.45	0.35	0.31	0.23

IOP Conf. Series: Materials Science and Engineering **842** (2020) 012010 doi:10.1088/1757-899X/842/1/012010

3.3. SEM observation result

Figure 2 shows the surface of the specimen on which the base material and the Ti thin film are deposited at a magnification of 33 times using a scanning electron microscope (SEM). It was observed that debris agglomerated due to the Ti thin film deposited (left). In the case of the base material (right), it was observed that no debris appeared. Many agglomerated debris were found, especially in 90' sputtered specimens. In the test process, it was confirmed that the thin film material adhered to the counterpart material. This is because the adhesion part of the two friction surfaces were sheared by the friction and peeled easily from the deposited thin film. In 90' sputtered specimens, it is thought that more debris was generated due to excessive deposition of thin films and poor bonding with the base material.

Table 4 is a table that summarizes the width and length of wear tracks for each specimen. The pattern is similar to the friction coefficient graph. The wear tracks of 30' and 60' sputtered specimens were narrower as the surface roughness decreased. The lower the roughness of the 90' sputtered specimen, the wider the track.

3.4. Wear loss measurement result

Figure 3 is a graph of the wear amount of each specimen. The 30' and 60' sputtered specimens showed a decrease in wear with surface roughness. In particular, the 7 and 8 specimens showed lower wear than the base metal. On the other hand, in 90' sputtered specimens, the lower the surface roughness value, the higher the wear.

Through this study, it was confirmed that the surface roughness and the deposition time of the thin film influence the wear resistance.

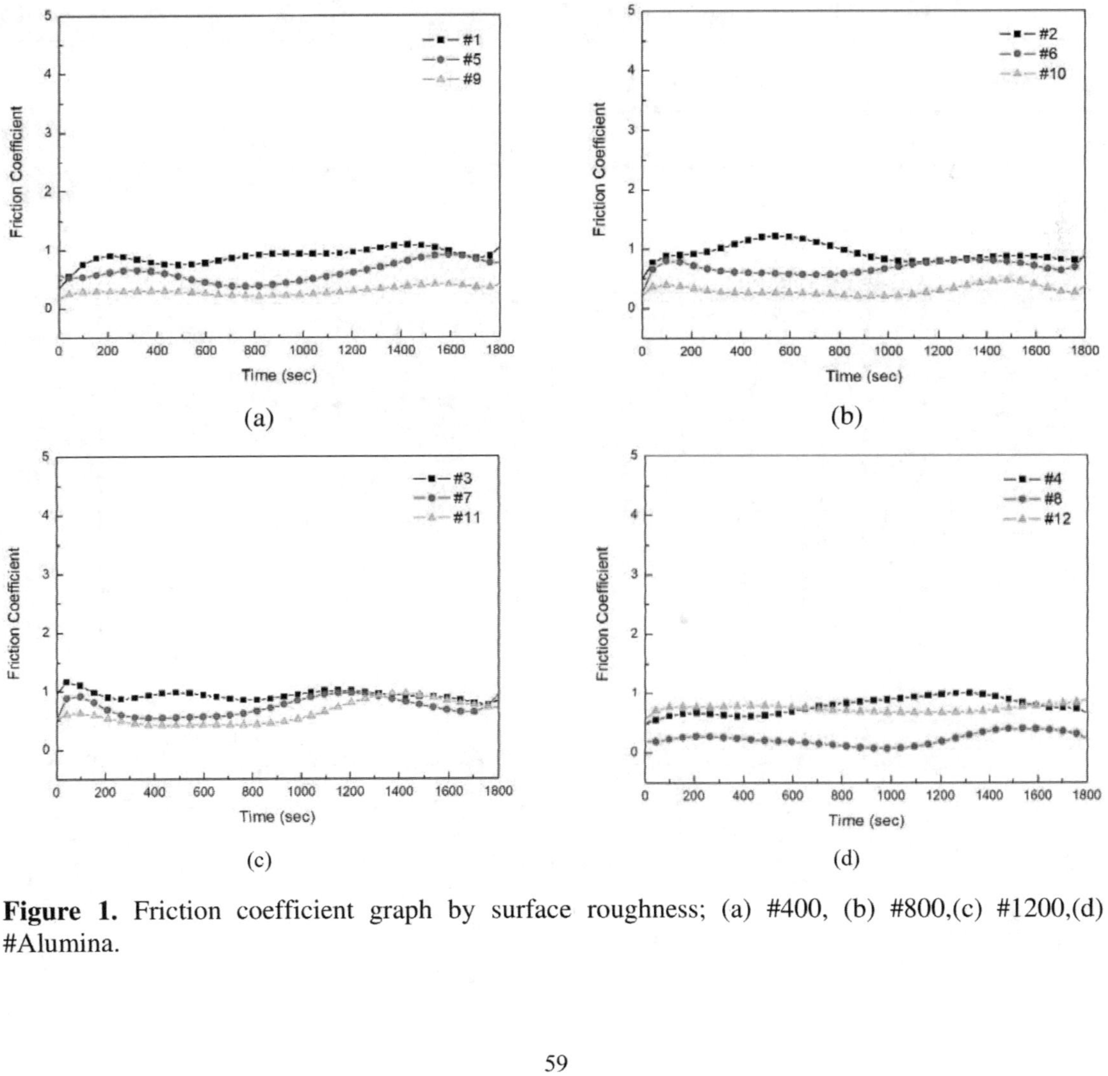

Figure 1. Friction coefficient graph by surface roughness; (a) #400, (b) #800,(c) #1200,(d) #Alumina.

Figure 2. Photograph of SEM wear track (x33).

Table 4. Average wear track width (mm).

1	2	3	4	5	6	7	8	9	10	11	12	13
1.601	1.502	1.386	1.475	1.287	1.284	1.002	0.804	0.84	0.846	0.97	0.951	0.814

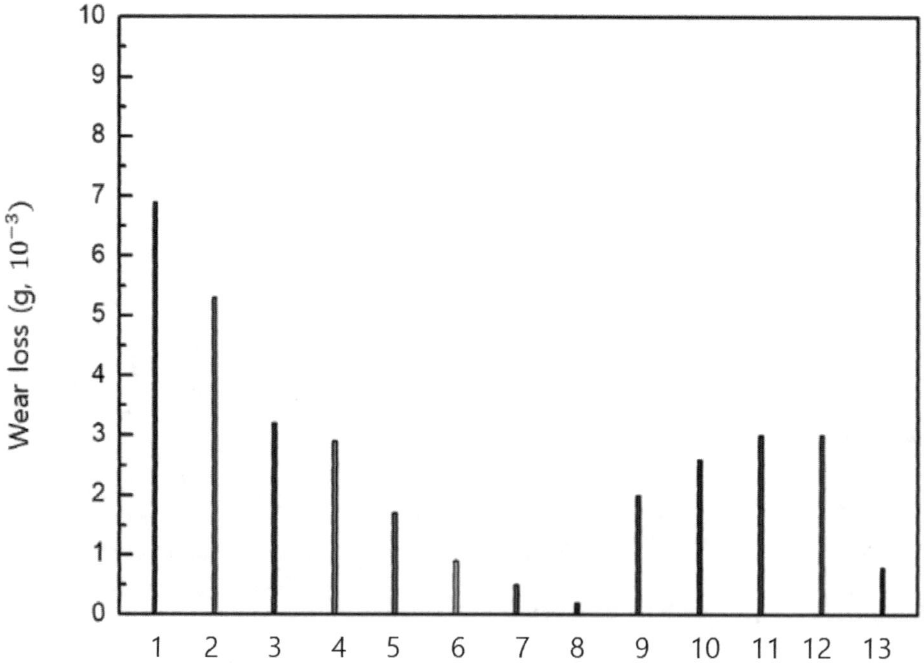

Figure 3. Wear loss graph.

4. Conclusion

In this study, we carried out wear tests to examine effects of surface roughness and film thickness of base material on wear properties.

(1) As a result of the analysis of wear characteristics, specimens with 30' and 60' sputtering tended to have a small friction coefficient, wear track width and wear as the deposition time increased. In contrast, the 90 'sputtered specimens tended to increase the coefficient of friction as the surface roughness decreased. The wear tracks showed debris. Particularly, more debris was observed in 90' sputtered specimens.

(2) Among all the specimens, 60' sputtered specimen (8) after polishing with alumina showed lower abrasion coefficient, narrower wear track, and less abrasion resistance than the base metal.

(3) In this experiment, it was confirmed that the adhered portions of the two friction surfaces were sheared by friction and peeled easily from the deposited thin film to show the appearance of adhesion wear.

Through this study, it was confirmed that the surface roughness and the thickness of the thin film influence the improvement of wear resistance. Based on these findings, the development of composite materials has economic and technical advantages and is worth investing in R & D. In addition, further experiments are planned based on the results of this study.

5. References

[1] E. S. Kim *A Study on the Evaluation of Wear-Resistance in Al 7075 Alloys* (KSMT, Korea, 2016), pp. 102-105.

[2] S. H. Park, W. J. Park, K. Y. Lee and S. C. Huh, *A Study on Wear Characteristics of Aluminium Alloy by Surface Hardening* (KSME, Korea, 2016), pp.45-50.

[3] C. B. In and S. S. Chun, *Wear Behavior of TiN Coatings Deposited by Plasma Assisted Chemical Vapor Deposition* (KJMR, Korea), vol. 3, no. 5, pp. 451-458.

[4] J. G. You, S. G. Kim, G. H. Lee and S. C. Kwon, *The Characterisitics of Cear-resistance Depending on Adhesion Strength for TiN Films Coated by DC Magnetron Sputtering* (thesis for a degree, PKNU, Korea, 2016)

[5] T. H. Kim, *Characteristic Study on Tribological Characteristic of DLC Film Deposited on Graphene Layer* (thesis for a degree, PSNU, Korea, 2013)

[6] J. H. Jang, S. H. Tak, Q. Jang, C. S. Huh and S. K. Lyu, *Effect of the Ti-series Coating on the Friction and Wear Characteristics of SCM415 Steel* (KSTLE, Korea) vol. 26, no. 3, pp. 162-166.

[7] M. S. Kim, J. H. Kho and S. H. Kim, *A Study on the Friction and Wear Characteristic of TiAlN and CrAlN Coating on the SKD61 Extrusion Mold Steel for 6xxx Aluminum Alloy* (J. Kor. Inst. Surf. Eng., Korea), vol. 43, no. 6, pp. 278-282.

Acknowledgement

This project was supported by the Basic Science Research Program through the National Research Foundation of Korea (NRF) funded by the Ministry of Science, ICT and future Planning (No. 2018R1A5A5A05022883).

And "Overseas order-linked aviation parts industry process technology development in 2019" of the Korea Institute for Advancement of Technology (KIAT), granted financial resource from the Ministry of Trade, Industry & Energy, Republic of Korea. (P0010339)

IOP Conf. Series: Materials Science and Engineering **842** (2020) 012011 doi:10.1088/1757-899X/842/1/012011

Luminescent Properties of Al_2O_3:$_x$Eu obtained by Rapid Combustion Synthesis

E. Viesca-Villanueva[2], D. Chavez[1,2,a], C. R. García[1,2,b], B. Esquivel[1], E. Martínez-Sánchez[3], C.M. Lopez-Badillo[2]

[1]Facultad de Ciencias Físico Matemáticas, Universidad Autónoma de Coahuila, Unidad Camporredondo, Edif. A, 25000, Saltillo, Coah., México.
[2]Programa de Posgrado de Doctorado en Ciencia y Tecnología de Materiales, CyTEMA, Facultad de Ciencias Químicas, Universidad Autónoma de Coahuila, 25280 Saltillo, Coahuila, México.
[3]CONACYT-Instituto Potosino de Ciencia y Tecnología, San Luis Potosí, México.

[a)] david_chavez@uadec.edu.mx, [b)] crodriguezgarcia@uadec.edu.mx

Abstract. In this work, $Al_2O_3{:}xEu^{3+}$ were synthesized by the combustion method. The atomic concentration percentage of Eu doping was varied (x = 0%, 1%, 3%, 5% and 7at%). Structural and morphological properties were analyzed by X-ray diffraction (XRD) and Scanning Electron Microscopy (SEM) techniques, respectively. The XRD patterns of the pure (x = 0%) and Eu-doped (x=1at%) show the single Al_2O_3 corundum phase. For the highest Eu-doped concentrations of x = 3%, x = 5% and x = 7at%, the Al_2O_3:xEu samples, present a mixture of phases of corundum (Al_2O_3) and $EuAlO_3$ (perovskite). SEM images show irregular grains with average sizes in the range of 0.1 – 78.8 μm. Photoluminescence (PL) analysis show that all Eu-doped Al_2O_3 samples exhibits intense red-emission. PL spectra show two main peaks at 615 nm ($^5D_0 \rightarrow {}^7F_2$) and 700 nm ($^5D_0 \rightarrow {}^7F_4$). The x = 3at% Eu-doped sample presents the highest intensity under 256 nm excitation. CIE coordinates were calculated from the PL spectra, values around x= 0.6 and y= 0.3 (red emission) were obtained from all samples. Due to its strong luminescence those Al_2O_3:xEu phosphors could be good candidates for components in solid state lighting.

1. Introduction

A challenge for nanoscience and nanotechnology is the development of new phosphor materials. The research is focused to obtain red, green and blue phosphors, due to these colors are classified as primary according to the RGB model proposed by Young-Helmholtz theory of trichromatic color vision [1].

Al_2O_3 alumina is a widely used low cost material, it has been used in abrasive ceramic applications and as a host lattice for rare-earths to produce luminescence [2, 3].

The europium is commonly used to produce the green and red emissions by doping some host crystalline lattices Eu^{2+} and Eu^{3+} ions, respectively. In the literature, $Al_2O_3{:}Eu^{3+}$ is reported as a good candidate to produce the red emission due to the f-f forbidden Eu^{3+} transitions. Those f-f transitions are excited by the Charge Transfer Band (CTB) of Eu-O and the excitation lines of Eu^{3+}, located at ~ 250 nm and 397 nm, respectively [4].

IOP Conf. Series: Materials Science and Engineering **842** (2020) 012011 doi:10.1088/1757-899X/842/1/012011

In 2008 Vinicius S.M. de Barros et al [5] worked on combustion synthesis to prepare Eu and Tb doped Al_2O_3 studying the thermoluminescent properties. In this research, a mixture of nitrates and urea were used for the combustion and annealed at 1000 °C, after that $\alpha - Al_2O_3$ was obtained.
In 2016 Yuya Onishi et al [6] worked on the solubility limit and luminescence properties of Eu doped Al_2O_3 powder. In this work, the synthesis was carried out mixing salts with urea and rare earths and sintered at 1200 °C in O_2 atmosphere, which increase the cost of the material production.
In 2017 Yue Hu et al synthesized $Al_2O_3{:}Eu^{3+}$ red-emitting spherical phosphors by hydrothermal method and analyzed their luminescence characteristics [7]. However, the hydrothermal method involves higher temperature, around 1300°C, and many days of synthesis, which provokes a most expensive producing method not being suitable for scalable production.
Based on the above, the aim of this work is to demonstrate that by using the Combustion Synthesis is possible to obtain $Al_2O_3{:}_xEu$ (x = 0%, 1%, 3%, 5% and 7at%) red-emitting phosphors in a rapid time of 5 min, with a subsequently post-annealing treatment for 1 hour in order to enhance its luminescence intensity.

2. Results and discussion

The Figure 1a shows the simulation of the Al_2O_3 corundum phase structure using Diamond V3.2k software, the white and blue spheres represent the Al and O atoms, respectively. The X-ray diffraction patterns of all samples with the different x doping concentrations (x=0%, 1%,3%, 5%, 7%) are shown in figure 1b. The PDF pattern is plotted in vertical lines below the XRD powder patterns. The x=0% label is associated to diffraction pattern of the Al_2O_3 undoped sample, that powder showed an XRD pattern which matches very well with the Al_2O_3 corundum phase, when x= 1% of Eu, the phase does not change with the augmentation of Eu^{3+}. A mixture of phases Al_2O_3 corundum and $EuAl_2O_3$ perovskite could be appreciated in x= 3%, 5% and 7%, the peaks corresponding to $EuAl_2O_3$ are marked with "*" according to the JCPDS 30-0012 and it has been already reported [8]. It could be asserted that in samples with less concentration of Eu^{3+} (x = 0 %, 1 % at) the sites of Al^{3+} are occupied by the Eu^{3+} ions while high concentration samples present the $EuAlO_3$ phase, thus is evident that Al_2O_3 has problems of solubility, the ionic radii of the 4-coordinate Al^{3+} ion is 0.53 Å while the Eu trends to present 6 and 8 coordinated structures and their ionic radii are of 1.31 Å and 1.2 Å, respectively. If we calculate consider that the ionic radii sizes of europium ions are more than the double that aluminum ions it is clear that the solubility will be poor.

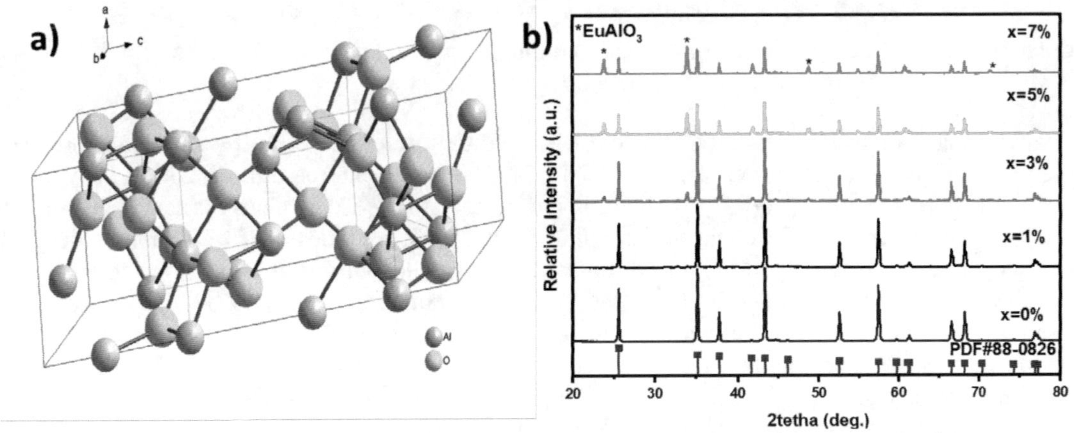

Figure 1.: a) Simulation of Al_2O_3 corundum phase. b) XRD Patterns of Al_2O_3:xEu samples.

The SEM images are shown in figure 2, all samples show irregular shapes, the sizes of the grains vary from 0.1 μm to 78.8 μm with an average size of 26.4 μm. All samples present coalesced particles, this is typical of aluminum oxides synthesized by combustion method as it was observed by [9] in $CaLa_{0.97}Eu_{0.03}Al_3O_7$. The sample with x=5% is not shown at the figure 2 but the images are much the same to all other samples. The augment of the micrographs was varied from 400x to 18,000x, all

IOP Conf. Series: Materials Science and Engineering **842** (2020) 012011 doi:10.1088/1757-899X/842/1/012011

grains shows irregular shapes and irregular agglomerates typical of some luminescent materials that tend to form groups and conglomerates [10].
Figure 3a shows the photoluminescence (PL) excitation spectra observing the emission wavelength at 615 nm. Figure 3b shows the emission spectra under 256 nm excitation. The highest PL intensity was achieved when the Al_2O_3 was doped with x=3.0 at % of Eu. All samples showed similar peaks profiles, the most intense peaks are located at 590 nm, 615 nm, 651 nm and 700 nm, those peaks are associated to the interatomic Eu^{3+} forbidden transitions: $^5D_0 \rightarrow ^7F_j$ transitions with j = 0, 1, 2, 3 and 4, respectively [9-12]. The transition corresponding to the peak located at 631 nm is due to an electric dipole transition for $\Delta j=\pm2$, the transition of the peak located at 590 nm is attributable to typical magnetic allowed transitions when $\Delta j= \pm1$.

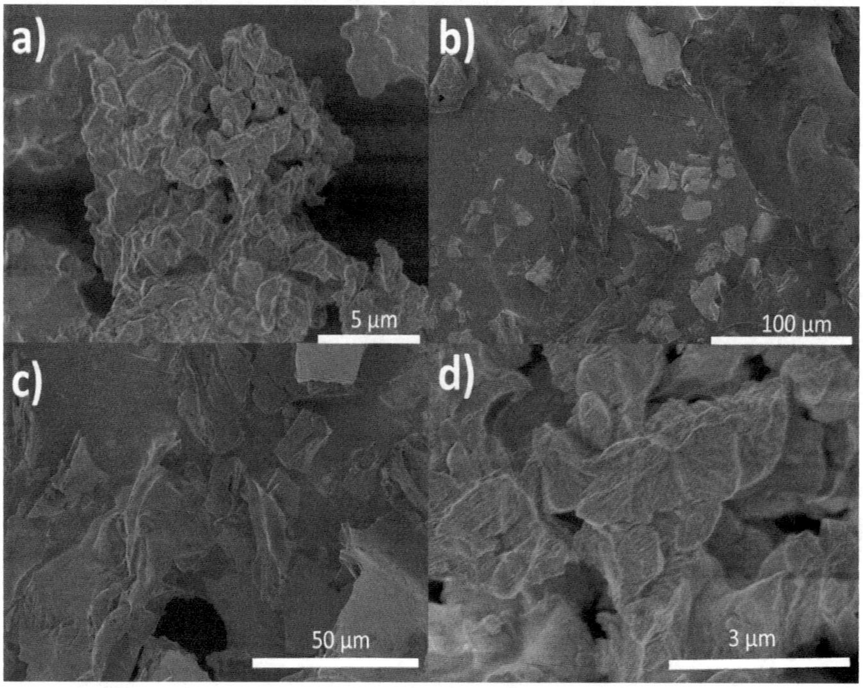

Figure 2. SEM micrographs of the Al_2O_3:xEu samples. a) x = 0.0%, b) x = 1.0% , c) x = 3.0% and d) x = 7.0%.

The peaks situated at 651 nm and 700 nm are originated by forbidden $^5D_0 \rightarrow ^7F_3$ transition and electric dipole allowed $^5D_0 \rightarrow ^7F_3$ transitions, respectively [11]. The band centered at peak 256 nm is very wide as it can be appreciated in figure 3a, the band is attributable to the Eu-O charge transfer process [10], this leads to be more suitable material for white light emitting diodes (WLEDs) applications. The peaks 361 nm, 380 nm and 397 nm are also reported by [9]. The splitting lines observed at the peaks at 590, 615, 651, and 700 nm (shown in figure 3b) can be attributed to the crystal field spectra of Eu^{3+} as it is reported in [12]. It is well known that the relative intensity of $^5D_0 \rightarrow ^7F_3$ transition and $^5D_0 \rightarrow ^7F_3$ transition robustly depends on the local symmetry of Eu^{3+} ions. Consequently, the $^5D_0 \rightarrow ^7F_1$ transition should be relatively strong as the Eu^{3+} ions occupy the inversion center sites [10], whereas the $^5D_0 \rightarrow ^7F_2$ transition is parity forbidden and must be very weak [9,10].

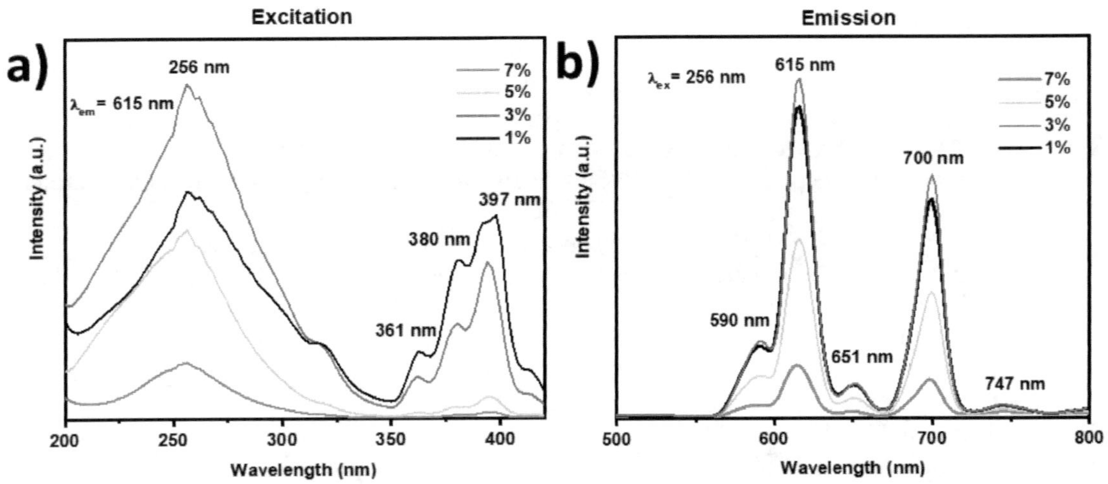

Figure 3. a) Excitation spectra of Al_2O_3: xEu.samples. b) Emission spectra of Al_2O_3: xEu.samples.

The CIE Coordinates diagram map of Al_2O_3: xEu samples are shown in figure 4, an inset table is added to facilitate and corroborate the points marked at the diagram. The color of the sample with x= 1.0% trends to be orange meanwhile the other samples with x = 3.0%, 5.0% and 7.0% are clearly red.

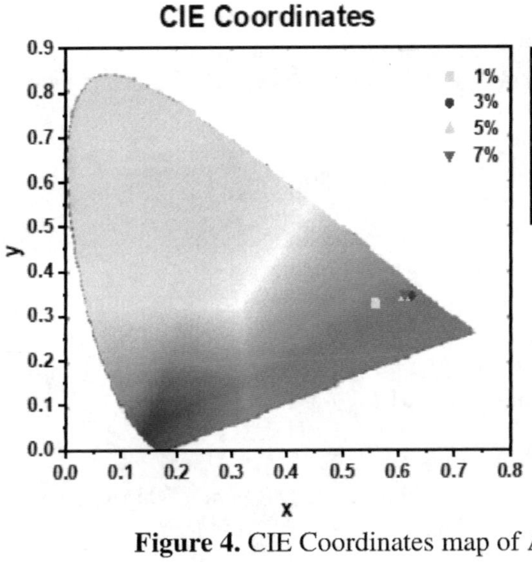

Eu x doping (%)	x	y
1	0.56	0.327
3	0.623	0.344
5	0.611	0.342
7	0.612	0.347

Figure 4. CIE Coordinates map of Al_2O_3: xEu.samples.

3. Conclusions

Combustion synthesis technique was useful to obtain the Al_2O_3:xEu phosphors, the materials were obtained by a rapid and efficient combustion method. The phases obtained were Al_2O_3 (corundum) and $EuAlO_3$ (perovskite). The SEM micrographs showed irregular shapes in all samples with variable sizes in the range of 0.1 – 78.8 μm. The most intense luminescence was presented by the x=3.0at% Eu-doped sample, a mixture of phases is presented in this Eu^{3+} concentration but it the luminescent processes were not inhibited. The excitation peak was located at 256 nm, and the most intense peaks were situated at 615 and 700 nm due to the ($^5D_0 \rightarrow {}^7F_2$) and ($^5D_0 \rightarrow {}^7F_4$) transitions, respectively. The CIE coordinates shown the changes in the color from orange to red when the Eu^{3+} concentration increases from x = 0.56, y = 0.327 to x = 0.612, y = 0.347. The Al_2O_3:xEu phosphors results suitable to be applied in components of solid-state lighting devices due to its ease and rapid obtention process.

4. References

[1] S. Natarajan, H. C. Bajaj, and R. J. Tayade, "Recent advances based on the synergetic effect of adsorption for removal of dyes from waste water using photocatalytic process," *J. Environ. Sci.*, vol. 65, pp. 201–222, 2017.

[2] M. R. Al-mamun, S. Kader, M. S. Islam, and M. Z. H. Khan, "Photocatalytic activity improvement and application of UV-TiO_2 photocatalysis in textile wastewater treatment: A review," vol. 7, no. July, 2019.

[3] D. Zhu and Q. Zhou, "Action and mechanism of semiconductor photocatalysis on degradation of organic pollutants in water treatment: A review," *Environ. Nanotechnology, Monit. Manag.*, vol. 12, no. July, p. 100255, 2019.

[4] K. Wetchakun, N. Wetchakun, and S. Sakulsermsuk, "An overview of solar / visible light-driven heterogeneous photocatalysis for water purification: TiO_2 - and ZnO-based photocatalysts used in suspension photoreactors," *J. Ind. Eng. Chem.*, vol. 71, pp. 19–49, 2019.

[5] M. A. Fox and M. T. Dulay, "Heterogeneous Photocatalysis," *Chem. Rev.*, pp. 341–357, 1993.

[6] M. R. Dhananjeyan, R. Annapoorani, S. Lakshmi, and R. Renganathan, "An investigation on TiO_2-assisted photo-oxidation of thymine," *J. Photochem. Photobiol.*, vol. 96, pp. 187–191, 1996.

[7] Y. Sun, S. Wang, and J. Zheng, "Biosynthesis of TiO_2 nanoparticles and their application for treatment of brain injury:An in-vitro toxicity study towards central nervous system," *J. Photochem. Photobiol. B Biol.*, vol. 194, no. 1277, pp. 1–5, 2019.

[8] V. Thiagarajan *et al.*, "Diminishing bioavailability and toxicity of P25 TiO_2 NPs during continuous exposure to marine algae Chlorella sp .," *Chemosphere*, vol. 233, pp. 363–372, 2019.

[9] X. Jin, X. Zhou, P. Sun, S. Lin, W. Cao, and Z. Li, "Photocatalytic degradation of nor fl oxacin using N-doped TiO_2: Optimization , mechanism , identi fi cation of intermediates and toxicity evaluation," *Chemosphere*, vol. 237, 2019.

[10] G. H. Da *et al.*, "Toxicity assessment of TiO_2 -MWCNT nanohybrid material with enhanced photocatalytic activity on Danio rerio (Zebra fi sh) embryos," *Ecotoxicol. Environ. Saf.*, vol. 165, no. March, pp. 136–143, 2018.

[11] Z. Khan, D. Shahwar, M. Khalil, Y. Ansari, and R. Chandel, "Toxicity assessment of anatase (TiO_2) nanoparticles: A pilot study on stress response alterations and DNA damage studies in Lens culinaris Medik," *Heliyon*, vol. 5, no. July, p. e02069, 2019.

[12] B. Yu, F. Meng, M. Wasim, R. Qin, and X. Liu, "Facile synthesis of AgNPs modi fi ed TiO_2 @ g-C_3N_4 heterojunction composites with enhanced photocatalytic activity under simulated sunlight," *Mater. Res. Bull.*, vol. 121, no. July 2019, 2020.

Acknowldegements

D. Chavez acknowledge to CONACYT-Mexico for the Doctorate scholarship 693166. C.R. Garcia acknowledge to Universidad Autónoma de Coahuila for support this research.

IOP Conf. Series: Materials Science and Engineering **842** (2020) 012012 doi:10.1088/1757-899X/842/1/012012

The Electronic and Magnetic Properties of Ultrathin γ-Fe_2O_3 Films

S Alraddadi

Department of Physics, Umm Alqura University, Makkah 24382, KSA

swraddadi@uqu.edu.sa

Abstract. In this work, iron oxide γ-Fe_2O_3 films with thickness of 10 nm were grown on a single crystalline MgO (001) substrate by molecular beam epitaxy using oxidation of an iron thin film at substrate temperature 250 °C. The crystal structure, electronic, and magnetic properties of the ultrathin epitaxial γ-Fe_2O_3 films were investigated. X-ray photoemission spectroscopy, Low energy electron diffraction, and X-ray diffraction confirmed that films were single crystalline quality with the same orientation as the MgO substrates and had only a single phase of γ-Fe_2O_3. Furthermore, the magnetization measurements at 300 K showed that the ultrathin films were ferromagnetic with a magnetization value of 270 emu cm^{-3}. These results show that it is possible to synthesize high-quality ultrathin γ-Fe_2O_3 films with good properties, which are promising as a spin filtering tunnel junction and for application in other spintronic devices, using the iron oxidation method.

1. Introduction

The studies and applications of iron oxide (γ-Fe_2O_3) thin films have progressed over the last decades due to their interesting electrical and magnetic properties. The iron–oxygen system has several oxides with different oxidation states and crystalline structures, including hematite (α-Fe_2O_3), maghemite (γ-Fe_2O_3), wustite (FeO) and magnetite (Fe_3O_4). Wustite and hematite are antiferromagnetic, whereas the other two oxides are ferromagnetic. γ-Fe_2O_3 is a ferromagnetic insulator, and thus may be a good candidate for developing room temperature spin filter devices. As a potential candidate for the spintronics applications, it is desirable to obtain high quality γ-Fe_2O_3 thin films. However, metastable γ-Fe_2O_3 is difficult to obtain and it can be easily transformed into antiferromagnetic α-Fe_2O_3 which is the equilibrium phase when heated above 350 °C [1,2]. To acquire γ-Fe_2O_3, a viable method may be the use of substrates with a slight lattice mismatch, such as MgO (001) substrates. Furthermore, the quality of the crystal structure of thin films is affected by several factors, including (i) the growth technique, (ii) substrate nature and crystallinity, (iii) substrate temperature, (iv) growth rate, (v) annealing temperature (vi) oxygen partial pressure and, heating rate. γ-Fe_2O_3 films have been prepared by different techniques, such as sputtering [3–5], pulsed laser deposition [6], atomic layer deposition [7], spray pyrolysis [2, 8], sol–gel deposition [9], electron-beam deposition [10], and chemical and metal organic deposition onto MgO substrates[1,11-17]. However, the quality of the films was not ideal, either with a decreased Curie temperature or a reduced magnetization. Therefore, in this work we present the synthesis of stoichiometric ultrathin γ-Fe_2O_3 films grown on a MgO (001) substrate by molecular beam epitaxy (MBE) using the oxidation method of iron thin films, and investigated the electronic and magnetic properties of the films. MBE deposition is an important technique for the deposition of oxide materials that can grow ultrathin films with very precise control over the thickness and stoichiometry. The γ-Fe_2O_3 films were characterized in-situ by X-ray photoemission spectroscopy

(XPS) and Low energy electron diffraction (LEED) as well as ex-situ by X-ray diffraction (XRD) and magnetic measurements.

2. Experiment

The epitaxial γ-Fe_2O_3 films were grown by molecular beam epitaxy (MBE) on single crystalline MgO (001) substrates. MBE was performed in an ultra-high vacuum (UHV) chamber with a base pressure of ~2.5×10^{-10} Torr. Before film deposition, the MgO (001) single crystal substrates were cleaned by acetone and methanol in ultrasonic baths followed by heating up to 300°C at $p(O_2) = 2\times10^{-7}$ Torr for 30 minutes in the UHV chamber. Afterward, the quality of the crystal and surface of the substrates were confirmed by the sharp low energy electron diffraction (LEED) pattern and X-ray photoemission spectroscopy (XPS).

γ-Fe_2O_3 thin films of 10 nm thickness were prepared using the oxidation method of iron thin films. For this purpose, pure iron rods were thermally evaporated at a deposition rate of 3.33 Å/min with substrate temperatures 250 °C, followed by oxidation of the iron film under $p(O_2) = 7\times10^{-7}$ Torr for 30 minutes at substrate temperature 250 °C. The oxygen was introduced to the chamber by a leak valve. The electronic and crystal structures of the films were studied by X-ray photoemission spectroscopy (XPS) (Mg-Kα), Low energy electron diffraction (LEED), and X-ray diffraction (XRD), respectively. The measurements of the magnetic properties were carried out using a superconducting quantum interference device (SQUID) from Quantum Design in a temperature range from 5 to 300 K.

3. Results and discussion

In order to identify the chemical state of the iron oxide in the epitaxial film, the Fe 2p core level was analyzed by XPS. The iron 2p level consists of two sublevels ($2p_{3/2}$ and $2p_{1/2}$) due to spin–orbit splitting, and the value of the binding energy of the Fe $2p_{3/2}$ level depends on the oxidation state of the iron. The binding energy of Fe $2p_{3/2}$ in metallic iron (Fe^0) is in the range of 706.7–707.4 eV [18–20, 21]. The binding energies of Fe $2p_{3/2}$ in the Fe^{2+} (FeO) and Fe^{3+} (Fe_2O_3) states are in the range of 709.3–710.0 eV [22,18–24], and 710.6–711.6 eV [18–20, 22–25], respectively. Moreover, the different phases of iron oxide (FeO, Fe_3O_4, and Fe_2O_3) can be distinguished by their characteristic satellite peaks at binding energies of 715–716 eV for FeO [18, 21] and 719.0–719.8 eV for Fe_2O_3 [18, 21-22, 24]. On the other hand, for stoichiometric Fe_3O_4, the presence of both satellite peaks makes the distinct satellite peaks unresolvable [18]. These satellites are used as fingerprints to identify the iron oxide phases [25]. Figure 1 shows the Fe 2p core-level XPS spectra obtained with Mg-Kα X-ray radiation for an epitaxial 10 nm iron film grown on single crystalline MgO (001) substrate before and after oxidation. The binding energy of the Fe 2p spectrum was calibrated by the C1s binding energy at 284.5 eV. Figure 1(a), shows the XPS Fe 2p spectra of an Fe thin film grown on a MgO (001) substrate under UHV with substrate temperatures 250 °C. The binding energy of the Fe $2p_{3/2}$ peak was located at 707 eV with a narrow full width at half maximum. The Fe 2p spectra studied here show no signs of a satellite structure of iron oxide, which can be taken as an indication of stoichiometric Fe metal. In figure 1(b, a) strong satellite peak is observed at 719 eV for the thin film after oxidation under $p(O_2) = 7\times10^{-7}$ Torr at 250 °C, indicating the presence of Fe^{3+} ions. This satellite behavior indicates a stoichiometry of Fe_2O_3, which is consistent with previous studies [18, 21-22, 24]. Furthermore, the Fe 2p iron oxide peaks are significantly shifted to higher binding energies and are broader than the peak of the pure metal. The XPS results indicate that the iron film is fully oxidized after exposure to oxygen at a pressure of $p(O_2) = 7\times10^{-7}$ Torr for 30 minutes at substrate temperature 250 °C [25-28].

The corresponding satellite structure can be easily separated to distinguish among FeO, Fe_3O_4, and Fe_2O_3; but cannot distinguish between α-Fe_2O_3 and γ-Fe_2O_3 because both of these iron oxides have nearly the same XPS spectrum. The surface crystal structure of the iron films after oxidation at $p(O_2)$ $= 7\times10^{-7}$ Torr was characterized by LEED and is displayed in figure 2(a). The film had a (1×1) surface structure, as indicated by the small (blue) solid square and has a single crystalline quality with the same orientation as the MgO (001) substrates. This observation of a clean and well-ordered surface structure is an indication of the formation of a high-quality γ- Fe_2O_3, which is consistent with previous reports [29]. However, LEED is a surface sensitive technique that can only characterize the three atomic layers of a film (≈ 10 Å), and the other phases of an iron oxide such as α-Fe_2O_3 might be present in a lower layer of the film. Therefore, we need to check the entire thickness of the film by XRD. The XRD spectra of the iron oxide film deposited on MgO (001) substrate under $p(O_2) = 7\times10^{-7}$ Torr for 30 minutes at substrate temperature 250 °C are shown in figure 2.

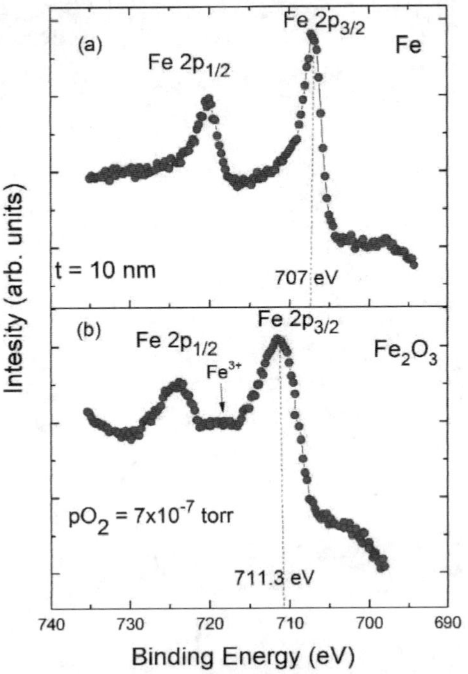

Figure 1. Fe 2p photoemission spectra obtained with Mg-Kα X-ray radiation for: (a) The epitaxial 10 nm of Fe thin film grown on MgO (001) with substrate temperatures 250°C, and (b) Iron oxide thin film prepared by oxidation of iron thin film under $p(O_2) = 7\times10^{-7}$ Torr for 30 minutes at substrate temperature 250 °C.

The film after oxidation has a single crystal structure with the same orientation as the substrates (001). The only observed peak for the film on the MgO substrate is located at the location of the substrate (002) peak, which indicates that the iron oxide film could be an epitaxial film.

Although the XRD measurements clearly demonstrated the epitaxial relation between the iron oxide film and the single crystalline MgO substrate, XRD could not be used to support the structural analysis that was obtained through the XPS and magnetization measurements due to the overlap of the broad and weak film peaks with the sharp and strong single crystalline MgO peaks[16,17].

In order to distinguish between α-Fe_2O_3 and γ-Fe_2O_3, magnetic measurements were performed. The measurements were made with a magnetic field of 500 Oe, applied in the (001) plane of the oxide film. The temperature dependence of the zero-field-cooled (ZFC) and field-cooled (FC) magnetization curves was measured for the film and are displayed in figure 3. The curve shows a typical temperature dependence of the magnetization of γ-Fe_2O_3. No deviation, such as the Morin transition for α-Fe_2O_3 or the Verwey transition for Fe_3O_4, occurred at 260 K and 124 K, respectively [30-33].

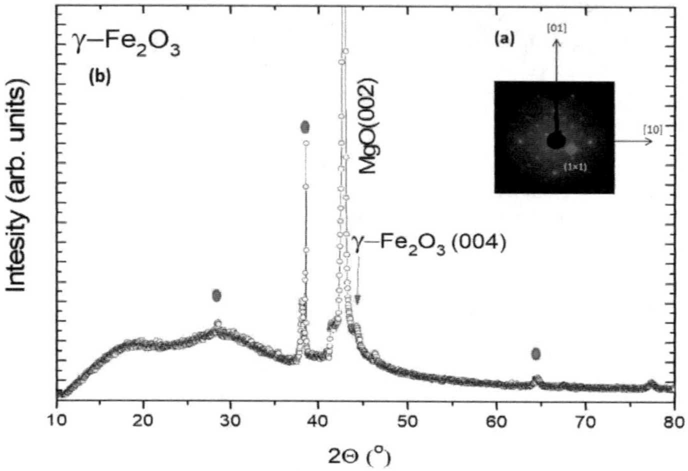

Figure 2. (a) The upper right inset displays the Low energy electron diffraction pattern of the epitaxial 10 nm of iron oxide thin film prepared under $p(O_2) = 7\times10^{-7}$ Torr for 30 minutes at 250 °C. The (1×1) surface structure for γ-Fe_2O_3 is indicated by the small (blue) solid square (b) XRD spectra of the iron oxide thin film prepared under $p(O_2) = 7\times10^{-7}$ Torr for 30 minutes at substrate temperature 250 °C. The red circle indicates the diffraction peak of MgO (001) substrate.

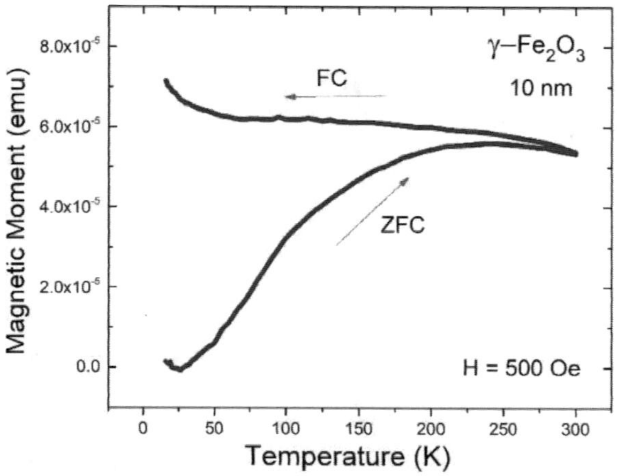

Figure 3. Total magnetic moment (emu) versus temperature (K) during heating, and subsequent cooling, for the epitaxial 10 nm iron oxide thin film prepared under $p(O_2) = 7\times10^{-7}$ Torr for 30 minutes at substrate temperature 250 °C.

Magnetic hysteresis loops for the 10 nm iron oxide films were measured at 300 K by applying the magnetic field in the (001) plane of the film. The films exhibit ferromagnetic behavior. Figure 4(a) shows the total measured magnetic moment for the film while figure 4(b) shows the film-only magnetization which was obtained by subtracting the MgO substrate contribution and dividing the value by the film volume. The saturation magnetization value at 300 K of the maghemite (γ-Fe_2O_3) films is 270 emu cm^{-3}, which is lower than the bulk value of γ-Fe_2O_3 (400 emu cm^{-3}). This result may be due to the anti phase boundary (APBs), because the density of APBs increases with decreasing thickness [34]. However, the value of the magnetic moment in our work is larger than that reported by *Mango et al.* and *Dung et al.* (~150 emu cm^{-3}) [16,33]. On the other hand, it was reported in earlier studies by *Huang et al.* and *Peng et al.* [17,35] that the magnetic moment of 10 nm thickness was close to that of the bulk. From these previous studies, it is clear that the magnetization property of a magnetite thin film strongly depends on the method of preparation, the nature of defects and the defect density. Based on the XPS and magnetic measurement results, we can conclude the formation of the stochiometric γ-Fe_2O_3 film.

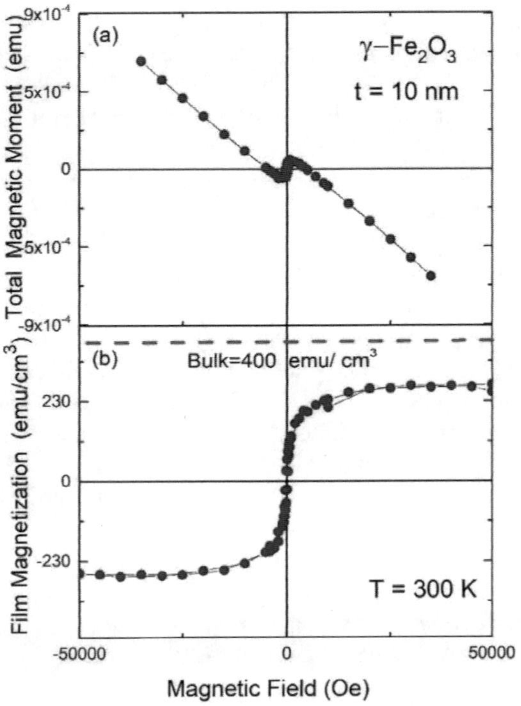

Figure 4. (a) Total magnetic moment (emu) versus magnetic field (Oe), hysteresis loops, obtained at 300 K for the 10 nm iron oxide thin film prepared under $p(O_2)=7\times10^{-7}$ Torr for 30 minutes at substrate temperature 250 °C (b) Film magnetization (emu/cm^3), obtained from the data in (a) corrected for the MgO substrate contribution.

4. Conclusions

In conclusion, we have successfully synthesized ultrathin γ-Fe_2O_3 films by MBE using the oxidation method of iron thin films. The structural and electronic characterization showed that the film fully oxidized after oxidation at $p(O_2) =7\times10^{-7}$ Torr and has a stoichiometric structure of γ-Fe_2O_3 with a crystallographic orientation the same as the substrate MgO (001). The magnetic measurements showed a ferromagnetic film with a magnetic moment value of ~270 emu cm^{-3}. These results indicate that a high quality ultrathin γ-Fe_2O_3 film can be achieved by oxidation of the iron thin film, and these films will be promising for spin filtering tunnel junctions and other spintronic applications. Furthermore, this growth method will be useful for increasing the quality of other metal oxide ultrathin films.

5. References

[1] Lee E T, Jang G E, Kim C K and Yoon D H 2001 Sens. *Actuator B- Chemical* **77**, 221-227.
[2] Dghoughi L, Elidrissi B, Bernede C, Addou M, Lamrani M A, Regragui M and Erguig H 2006 *Appl. Surf. Science* **253**,1823-1829.
[3] Stenberg T, Vuoristo P, Keränen J, Mäntylä T, Büchler M, Virtanen S, Schmuki P and Böhni H 1998 *Thin Solid Films* **312**, 46-60.
[4] Peng Y, Park C and Laughlin D E 2003 *J. Appl. Physics* **93**, 7957-7959.
[5] Miller E L, Paluselli D, Marsen B and Rocheleau R E 2004 *Thin Solid Films* **466**, 307-313.
[6] Shima M, Tepper T and Ross C A 2002 *J. Appl. Physics* **91**, 7920-7922.
[7] Lie M, Fjellvåg H and Kjekshus A 2005 *Thin Solid Films* **488**, 74-81.
[8] Ma T Y and Lee I C 2004 *J. Mater Sci: Mater Electronics* **15**(12), 775-780.

[9] Redel E, Mirtchev P, Huai C, Petrov S and Ozin G A 2011 *ACS nano* **5**, 2861-2869.
[10] Jain S, Adeyeye A O, Chan S Y and Boothroyd C B 2004 *J. Phys. D: Appl. Physics* **37**, 2720.
[12] Pal B, and Sharon M 2000 *Thin Solid Films* **379**, 83-88.
[12] Maruyama T and Kanagawa T 1996 *Society* **143**, 1675-1677.
[13] Shalini K, Subbanna G N, Chandrasekaran S and Shivashankar S A 2003 *Thin Solid Films* **424**, 56-60.
[14] Pflitsch C, Viefhaus D, Bergmann U, Kravets V, Nienhaus H and Atakan B 2006 *J. Electrochem. Society* **153**, 546-550.
[15] Singh M K, Yang Y and Takoudis C G 2008 *J. Electrochem. Society* **155**, 618-623.
[16] Manago T, Sonobe T, Yamaguchi I and Takano F 2010 *J. Appl. Physics* **107**, 053908.
[17] Huang X L, Yang Y and Ding J 2013 *Acta materialia* **61**, 548-557.
[18] Kuivila C S, Butt J B and Stair P C 1988 *Appl. Surf. Science* **32**, 99-121.
[19] Stambouli V, Palacio C, Mathieu H J and Landolt D 1993 *Appl. Surf. Science* **70**, 240-244.
[20] Graat P and Somers M A 1998 *Surf Interface Analysis* **26**, 773-782.
[21] Corneille J S, He J W and Goodman D W1995 *Surf. Science* **338**, 211-224.
[22] Gao Y, Kim Y J, Chambers S A and Bai G 1997 *J Vac Sci Technology* A **15**, 332-339.
[23] Vurens G H, Salmeron M, and Somorjai G A 1988 *Surf. Science* **201**, 129-144.
[24] Schedel-Niedrig T, Weiss W and Schlögl R 1985 *Phys. Review* B **52**, 17449.
[25] Fujii T, De Groot F M F, Sawatzky G A, Voogt F C, Hibma T and Okada K 1999 *Phys. Review* B **59**, 3195.
[26] Barbieri A, Weiss W, Van Hove M A and Somorjai G A 1994 *Surf. Science* **302**, 259-279.
[27] Moussy J B 2013 *J. Phys. D: Appl. Physics* **46**, 143001.
[28] Karunakaran C and Senthilvelan S 2006 *Electrochem. Communications* **8**, 95-101.
[29] Chambers S A and Joyce S A 1999 *Surf. Science* 420, 111-122.
[30] Morin F J 1950 Phys. Review **78**, 819.
[31] Suber L, Santiago A G, Fiorani D, Imperatori P, Testa A M, Angiolini M and Dormann J L 1998 *Appl. organom. chemistry***12**, 347-351.
[32] Verwey E J W 1939 *Nature* **144**, 327-328.
[33] Dung D D, Feng W, Van Thiet D, Kim Y and Cho S 2015 *Mater. Letters* **161**, 343-347.
[34] Hibma T, Voogt F C, Niesen L, van der Heijden P A A, de Jonge W J M, Donkers J J T M and Van der Zaag P J 1999 *J. App. Physics* **85**, 5291-5293.
[35] Li P, Jiang E Y and Bai H L 2011 *J. Phys. D: Appl. Physics* **44**,075003.

Acknowledgments
The author would like to acknowledge Prof. William Hines for experimental support and useful dissection.

IOP Conf. Series: Materials Science and Engineering **842** (2020) 012013 doi:10.1088/1757-899X/842/1/012013

A Study on the Thermal Properties of Thermal Grease with Copper Nanopowders

HaNeul Kang[1], HyunJi Kim[1], SangHyoek Park[1], JinHo Yang[2], Sunchul Huh[3*]

[1]Dept. of Energy and Mech. Eng. Graduate school of GNU, Tongyeong-si, Korea
[2]Dept. of Computer Aided Machinery Iksan Campus of Korea Polytechnic Iksan-si, Korea
[3]Dept. of Energy and Mech. Eng. Of GNU, Tongyeong-si, Korea

[*]schuh@gnu.ac.kr

Abstract. Electronic devices and main boards have a tendency to increase the internal heat radiation performance for excellent thermal conductive materials, as necessary for miniaturization. In this study, Nano-thermal grease was prepared by mixing copper Nano-powder into thermal grease, one of the types of thermal conductive materials, by volume percentage. The thermal conductivity was measured and analyzed. As a result, the thermal conductivity was excellent in the order of Case 5, Case 4, Case 3, Case 2, Case 1. However, when the copper powder of Case 5 or more samples was added, stirring did not proceed smoothly due to the high viscosity. This is considered to be because the maximum capacity of the thermal grease was exceeded.

1. Introduction

With the need for producing devices and motherboards that are smaller in this modern-day, the need for thermal conductive materials that have excellent internal heat dissipation is increasing [1-3]. Therefore, there is a need for research on materials that possess higher performance heat transfer and resistance. Thermal management is directly related to the performance and efficiency of the equipment [4-6]. Thermal interface materials (TIM) are essential in the improvement of heat dissipation performance of electronic products with heat generation [7-8]. In particular, it is applied to the interface between the heat spreader and sink of the CPU or GPU and used to reduce the thermal contact resistance by removing the air layer present at the interface. In this study, we prepared a Nano-thermal grease mixed with silicone oil to the copper. Nano-powder was used as a heat transfer medium. In addition, the thermal properties of the prepared Nano-thermal grease was measured to compare the performance.

2. Experimental equipment and methods

Copper powder was produced using pulsed wire evaporation (PWE), one of the Nano-powder preparation methods. PWE is a method of producing Nano-sized powder by evaporation and condensation processes after subliming a certain length of metal wire by supplying high density electrical energy to the metal wire within a short period of time (0.0001 s). PWE may control the size of the powder by adjusting the intensity of the voltage applied to the metal wire [9]. The larger the voltage applied, the smaller the powder size, but the uneven size. Agglomeration of liquid particles other than gases during coagulation produces relatively large particles, whereas coagulation of semisolid particles that do not form a complete solid state produces coarse particles. Therefore, it is necessary to apply an appropriate voltage suitable for the diameter and length of the wire [10]. Figure 1 shows Equipment used to manufacture Cu nano powder.

Table 1 shows the production conditions of the copper Nano-powder. Powders of relatively constant sizes were prepared using appropriate voltages that matched the physical properties of copper. When the powder was prepared by the PWE equipment, the inside of the chamber was filled with argon gas to prevent the powder from being oxidized, and the powder was stored and treated by passivation in a sealed state for about 2 weeks after the powder was manufactured. For copper wire, the proper voltage is 4500V.

Figure 2 shows the process of manufacturing Nano-thermal grease. Each sample was mixed with a certain proportion of the powder and silicon oil according to the volume percentage based on the total amount of 100cc as shown in table 2. During the manufacture of the thermal grease, the copper Nano-powder was not uniformly mixed due to the high viscosity of the silicon oil. The mixture was stirred for 30 minutes at a speed of 300 rpm by using a heating plate at a high temperature of about 100 ° C.

(a)

(b)

Figure 1. Equipment used to manufacture nano powder: (a) Photograph of PWE equipment (b) Photograph of copper metal wire used in PWE equipment.

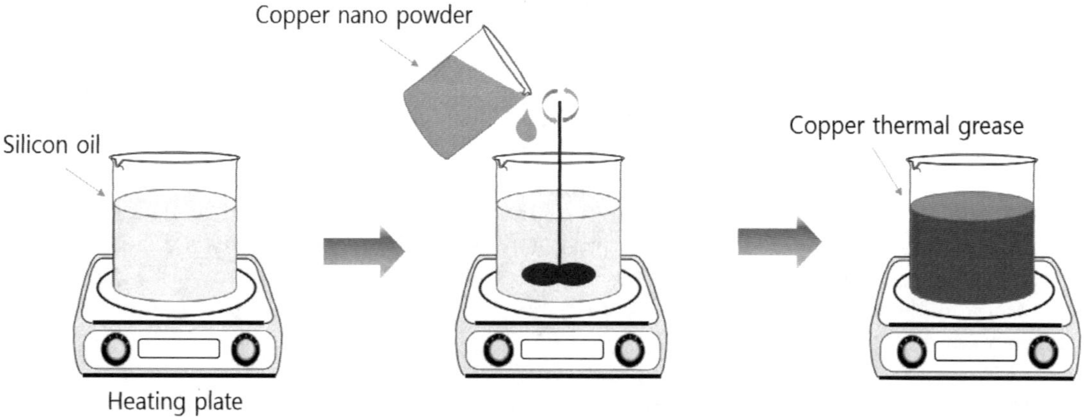

Figure 2. Manufacture process of nano thermal grease.

Table 1. Manufacturing conditions of copper nano powder.

Material	Voltage [V]	Number of Time	Wire Diameter [mm]	Wire Diameter [mm]
Copper	4500	2000	0.2	32

IOP Conf. Series: Materials Science and Engineering **842** (2020) 012013 doi:10.1088/1757-899X/842/1/012013

Table 2. Manufacturing conditions of each case of thermal grease

Case	Thermal grease [cc]	S/O [cc]	Vol%	Cu [g]
1	100	100	0	0
2		95	5	44.7
3		90	10	89.4
4		85	15	134.1
5		80	20	178.8

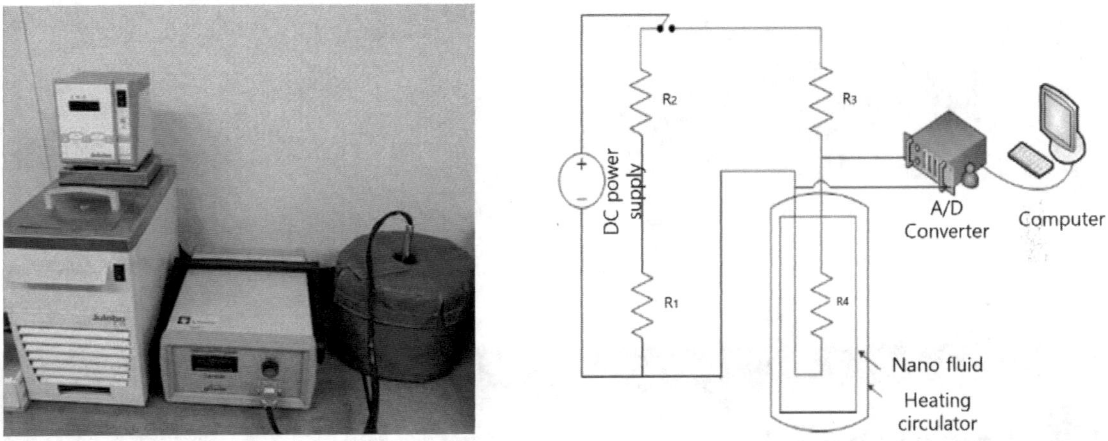

(a) (b)

Figure 3. Thermal conductivity measuring equipment: (a) Photograph of thermal conductivity meter (b) Schematic diagram of hot-wire apparatus.

3. Measuring equipment

Figure 3 is a photograph of the thermal conductivity measurement equipment used in the study, and is a schematic of the experimental setup of a hot-wire system. Thermal conductivity measuring equipment is a hot-wire method that has been widely used to measure the thermal conductivity of a fluid with high accuracy.

Based on the calculation of the temperature transport field around the metal wires inside, the electrical conductivity of the fluid is measured using an electric heating element and a resistance thermometer. It is a principle to measure. This method was first introduced by Nagaska and Nagashima et al. [11]. in 1981 and this hot-wire method has the advantage of being able to measure by minimizing the effects of natural convection and relatively fast measurement time [12-13].

4. Results

Figure 4 shows the SEI observed with FE-SEM of the shape and size of Cu powder prepared using PWE. During FE-SEM imaging, platinum coating pretreatment was performed and observed at 100,0kx magnification with a voltage of 15.0kx. And then, it can be seen that most of the powders are uniform and spherical in size with a radius of 100 nm. And a graph of EDS (Energy Dispersive Spectrometer) component detection analysis using FE-SEM. Of the components detected, C is assumed to be the carbon tape used for sample fixing, and O is assumed to be detected as the silicone oil and the oxidation reaction component. It can be confirmed that no component other than Cu is detected in the prepared powder. Figure 5 shows the SEI observed by the FE-SEM of the shape and size of Cu thermal grease prepared using PWE, a graph of the component analysis of the energy dispersive spectrometer (EDS) using the FE-SEM. Out of the components detected, C was assumed to be the carbon tape used for sample fixation, and O was assumed to be detected as the silicone oil and oxidation reaction component. It can be seen that no components other than Cu and Si were detected in the powders produced. And figure 6 shows the comparison of thermal conductivity.

(a)

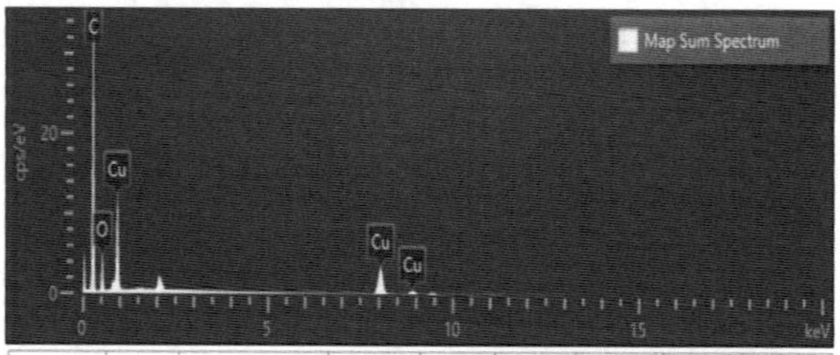

Element	Line Type	Apparent Concentration	k Ratio	Wt%	Wt% Sigma	Atomic %	Standard Label	Factory Standard
C	K series	12.05	0.12051	69.76	0.13	84.71	C Vit	Yes
O	K series	2.94	0.00991	12.23	0.10	11.15	SiO2	Yes
Cu	K series	7.97	0.07969	18.00	0.10	4.13	Cu	Yes
Total:				100.0		100.0		

(b)

Figure 4. FE- SEM Image of Cu nano powder manufactured using PWE equipment: (a) A SEI of Cu nano powder (b) EDS spectrum of Cu nano powder.

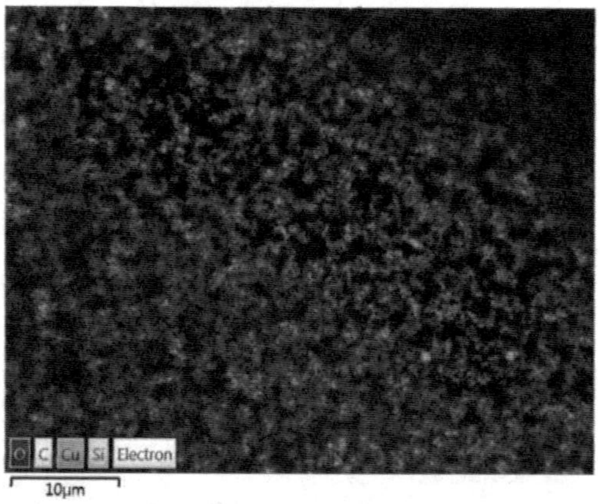

(a)

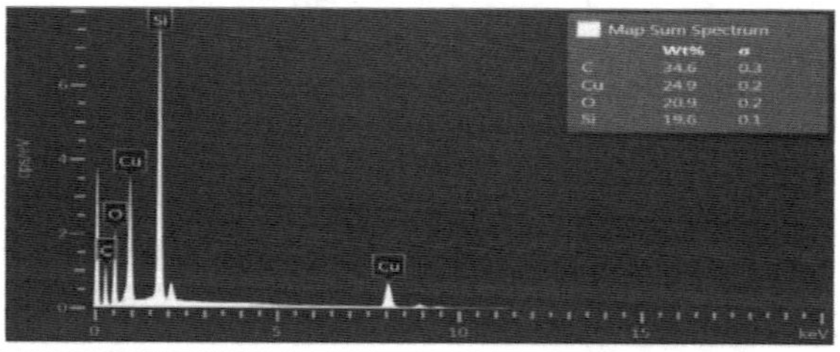

Element	Line Type	Apparent Concentration	k Ratio	Wt%	Wt% Sigma	Atomic %	Standard Label	Factory Standard
C	K series	1.03	0.01029	34.58	0.32	54.58	C Vit	Yes
O	K series	3.44	0.01159	20.88	0.18	24.74	SiO2	Yes
Si	K series	5.48	0.04339	19.63	0.12	13.25	SiO2	Yes
Cu	L series	3.94	0.03941	24.92	0.18	7.43	Cu	Yes

(b)

Figure 5. Image of EDS taken using FE-SEM: (a) A BEI of Cu nano thermal grease (b) EDS spectrum of Cu nano thermal grease.

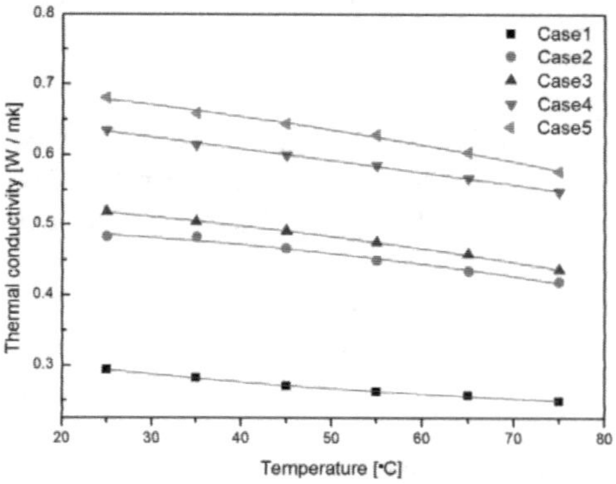

Figure 6. Graph of thermal conductivity measurement results for each sample.

5. Conclusions

In this study, Nano-thermal grease was prepared by mixing copper Nano-powder prepared by the electric line explosion method with silicone oil. Thermal conductivity was measured and analyzed. The conclusion is as follows.

- After BEI observation of Nano-thermal grease, Nano-powders are not agglomerated, and are evenly distributed.
- As a result of the thermal conductivity measurement, thermal conductivity was excellent in the order of Case 5, Case 4, Case 3, Case 2, and Case 1.
- It is considered that there may be an error due to air bubbles formed between samples when measuring the thermal conductivity, and further analysis will be performed after removing bubbles using ultra sonic.
- 30 vol% thermal conductivity is not manufactured due to the cohesion and high viscosity of the powder.

6. References

[1] Moore, A.L. and Shi, L., “Emerging challenges and materials for thermal management of electronics” Materials today, vol. 17, no. 4, 2014, pp. 163-174.

[2] Sarvar, F., Whalley, D. C. and Conway, P. P., Thermal interface materials-A review of the state of the art. In 2006 1st Electronic Systemintegration Technology Conference, Vol. 2, 2006, pp. 1292-1302. IEEE.

[3] Razeeb, K.M., Dalton, E., Cross, G.L.W. and Robinson, A.J., Present and future thermal interface materials for electronic devices. International Materials Reviews, 63(1), 2018, pp.1-21.

[4] Sartre, V. and Lallemand, M. Enhancement of thermal contact conductance for electronic systems. Applied thermal engineering, 21(2), 2001, pp. 221-235.

[5] Prasher, R., Thermal interface materials: historical perspective, status, and future directions. IEEE, 94(8), 2006, pp.1571-1586.

[6] Shahil, K. M., and Balandin, A. A., Thermal properties of graphene and multilayer graphene: Applications in thermal interface materials. Solid State Communications, 152(15), 2012, pp 1331-1340.

[7] Chung, D. D. L., Thermal interface materials. Journal of Materials Engineering and Performance, 10(1), 2001, pp. 56-59.

[8] Kreupl, F., Graham, A. P., Duesberg, G. S., Steinhögl, W., Liebau, M., Unger, E. and Hönlein, W. Carbon nanotubes in interconnect applications. Microelectronic Engineering, 64(1-4), 2002, pp. 399-408.

[9] Lee, G.H., Park, J.H., Rhee, C.K. and Kim, W.W., Fabrication of Al nano powders by pulsed wire evaporation (PWE) method. Journal of Industrial and Engineering Chemistry, 9(1), 2003, pp.71-75.

[10] Yılmaz, F., Lee, D.J., Song, J.W., Hong, H.S., Son, H.T., Yoon, J.S. and Hong, S.J., Fabrication of cobalt nano-particles by pulsed wire evaporation method in nitrogen atmosphere. Powder technology, 235, 2013, pp.1047-1052.

[11] Y.Nagasaka and A. Nagashima., Simultaneous measurement of the thermal conductivity and the thermal diffusivity of liquids by the transient hot - wire method. Riview of Scientific Instruments 52, pp. 229.

[12] Peñas, J. R. V., Ortiz de Zarate, J. M. and Khayet, M. Measurement of the thermal conductivity of nanofluids by the multicurrent hot-wire method. Journal of Applied Physics, 104(4), 2008, 044314.

[13] Junsun Hwan, Tae Hee Han, Performance-aware Dynamic Thermal Management by Adaptive Vertical Throttling in 3D Network-on-Chip. IEIE, 51(7), 2014, pp. 103-110.

Acknowledgments
This project was supported by the Basic Science Research Program through the National Research Foundation of Korea (NRF) funded by the Ministry of Science, ICT and future Planning (No. 2018R1A5A5A05022883).

IOP Conf. Series: Materials Science and Engineering **842** (2020) 012014 doi:10.1088/1757-899X/842/1/012014

Photocatalytic Evaluation of Bi-doped TiO/TiO_2 under Solar Irradiation

D. Chavez[1,2,a], C. R. Garcíal[2,b], J. Oliva[3], M.G. Godina Cubillo[1], R.C. Torres-Ramírez[1], C.M. Lopez-Badillo[2]

[1]Facultad de Ciencias Físico Matemáticas, Universidad Autónoma de Coahuila, Unidad Camporredondo, Edif. A, 25000, Saltillo, Coah., México.
[2]Programa de Posgrado de Doctorado en Ciencia y Tecnología de Materiales, CyTEMA, Facultad de Ciencias Químicas, Universidad Autónoma de Coahuila, 25280 Saltillo, Coahuila, México.
[3]CONACYT-Instituto Potosino de Ciencia y Tecnología, San Luis Potosí, México.

a) david_chavezc@uadec.edu.mx, b) crodriguezgarcia@uadec.edu.mx

Abstract. Abstract. The photocatalytic properties of TiO2 have been studied in the last decade due to its a high capacity to degrade organic compounds which dangerous for the human health. In order to increase its degradation efficiency, we propose the doping of TiO_2 with Bi dopant using concentrations from 0 to 5 at%. A simple and fast (40 min) combustion method was used to synthesize Bi doped TiO_2. According to the X-ray diffraction patterns, we obtained a mixture of TiO/TiO_2 phases with Bi dopant concentrations below 0.5 at.%. Above this value, the Bi7.68O12.16Ti0.32 compound was obtained. The morphological studies by Scanning electron microscopy demonstrated that a porous material is obtained for Bi concentrations below 0.5% but highly coalesced particles appeared for higher concentrations. The Bi doped materials were used to degrade Methylene blue dye under solar exposure and found a maximum degradation percentage of 96% for the sample doped with 0.1 at% of Bi.

1. Introduction

The water contamination problem has increased during the last years due to the use of dyes, herbicides and phenol derivatives in industrial processes, which are difficult to degrade. These contaminants are found in rivers and water effluents produced by the textile industry, fertilizer industry etc. [1]. Photocatalysis is considered a reliable alternative for the degradation of various organic and inorganic compounds in industrial wastewater through oxidation and reduction reactions. In this process, a photocatalyst is irradiated by light with energy equal or greater than its band gap [2–4], it absorbs the light and electrons are excited from the valence band (VB) to the conduction band (CB), which produces electron-hole (e-h) pairs. These ones react with water to produce OH- and O^{-2} radicals, which in turn, attack the dye or pharmaceutical contaminant, producing its degradation [5].
The industrial factories are looking for innocuous photocatalysts that can be activated with sunlight [6], [7]. In this sense, TiO_2 (Titania) photocatalyst could be an option because previous works demonstrated that it can degrade dyes and organic contaminants under Ultraviolet (UV) irradiation (λ < 385 nm) and this radiation could come from the sunlight. The Bi has been introduced as dopant in $_{TiO2}$ host to enhance its degradation efficiency, obtaining total degradation times of 30-180 min. An advantage of Bi doped materials such as TiO_2:Bi is their easy preparation using simple methods such

as sol-gel [8], [9]. Due to the advantages of Bi to enhance the photocatalytic degradation of dyes and the capacity of titania to absorb the UV component of the sunlight. We doped TiO_2 with Bi and evaluated its performance for the photocatalytic degradation of MB under sunlight irradiation.

2. Experimental

2.1. Combustion Synthesis of Bi-doped TiO_2 samples

The reagents were purchased from Sigma Aldrich and used as received. In a typical synthesis, Titanium isopropoxide [$C_{12}H_{28}O_4Ti$ (99.999%), Sigma Aldrich], bismuth nitrate pentahydrate [$Bi(NO_3)_2$ $5H_2O$ (98.0%) Sigma Aldrich], and glycine as fuel [$C_2H_5NO_2$ (99.0%), Faga Lab] were dissolved in a quartz beaker with 20 ml of distilled water under continuous stirring at 500 rpm. After this, a transparent blend was observed. Next, the beaker was introduced into a preheated furnace at 350°C to achieve the combustion process. After 10 minutes, a gray crystalline foam was produced from the combustion. Finally, the gray foam was grinded to obtain a photocatalyst powder. Several Bi-doped TiO_2 samples were synthesized with Bi dopant concentrations of x=0, 0.1, 0.3 and 0.5 at.%.
Photocatalysis experiments
The photocatalytic experiments were achieved for each Bi-doped TiO2 sample as follows: 30 mg of photocatalysts was added in 100 ml of aqueous methylene blue (MB) solution (25 ppm initial concentration). The MB was purchased from Sigma Aldrich with a 99.99% of purity. Immediately, the beaker was placed into a dark room for 1 h to allow adsorption of MB molecules at the surface of the photocatalyst. Subsequently, the solution was exposed to natural sunlight. Next, aliquots were extracted in regular periods of time for 240 minutes and their absorbance was measured using a UNICO SQ4802 UV-Vis spectrophotometer in the range of 200-800 nm. The MB degradation was observed by monitoring the decrease of the absorption peak at 665 nm, which is attributed to the MB dye. Thus, the degradation percentage of MB was calculated by:

$$\text{Degradation}=[(A_t-A_o)/A_o]*100\% \quad , \qquad (1)$$

where A_t is the absorbance at the time "t" and A_0 is the initial absorbance of the initial MB solution. In addition, the solar irradiance was monitored by using a Davis photodiode solarimeter detector each 30 minutes and the maximum value of the irradiance during the experiments was: 958 ± 10 W/m^2. The solar photocatalysis experiments were made at the astronomic observatory of Autonomous University of Coahuila in Saltillo México, which coordinates are Latitude: 25° 26' 0″ North Longitude: 101° 0′,0'' West, Height: 1581 m. in a sunny day starting at 12:14 h and finishing at 16:15 h.

3. Results and discussion

The X-ray diffraction patterns of TiO_2:Bi samples are shown in figure 1. Two phases appear in x = 0.0at%, 0.1at% and 0.3at% Bi-doped samples, that, this, TiO and TiO_2 (marked with α letter and PDF #96-900-8750). The sample with 0.5 at% of Bi dopant showed another phase, corresponding to the Bismuth titanium oxide: $Bi_{7.68}O_{12.16}Ti_{0.32}$ with PDF #96-200-5006 and labeled with β in figure 1. Thus, the increase of the Bi content provokes the phase change from TiO_2 – TiO to $Bi_{7.68}O_{12.16}Ti_{0.32}$.

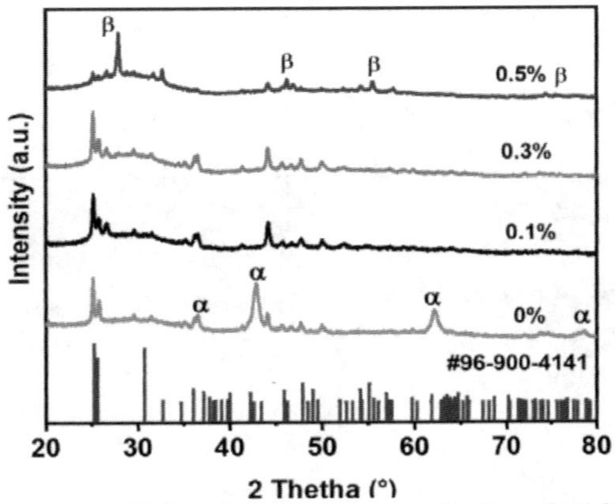

Figure 1. a) X-ray diffraction patterns of Bi-doped TiO_2 samples.

The SEM micrographs of Bi-doped TiO_2 samples are shown in figure 2. Figure 2a shows the typical porous morphology of undoped TiO_2 where grains with sizes from 0.2 to 1 μm are observed. The sample corresponding to x = 0.1at% is depicted in figure 1b, the morphology is similar, but it has bigger grains with sizes from 0.25 μm to 1.2 μm. The SEM images show coalesced grains in the sample with 0.3 at% of Bi dopant, see figure 1c. It shows irregular grains without the same porosity than the samples above and sizes from 0.6 μm to 6 μm. The most notorious change occurs in the sample with x=0.5at% in figure 1d, where conglomerated particles with sizes in the range of 1 μm to 15 μm are observed. The porosity was drastically reduced, and the particles have irregular shape.

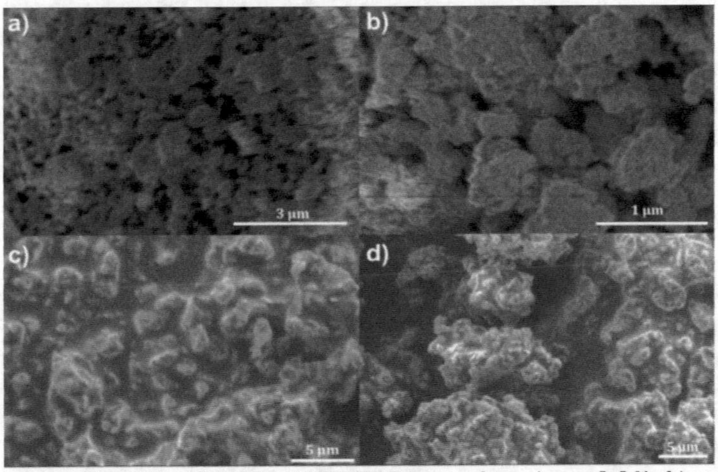

Figure 2. SEM micrographs of the Bi-doped TiO/TiO_2 samples: a) x = 0.0%, b) x = 0.1% and c) x = 0.3% as well as for $Bi_{7.68}O_{12.16}Ti_{0.32}$ (x = 0.5%).

Figure 3a shows the absorbance spectra obtained from the MB solution with the 0.1 at% photocatalyst, these spectra were obtained each 30 minutes for 4 hours. The absorption peak of MB is located at 665 nm. The solution changes its coloration from blue to transparent liquid after 240 min, as observed in the inset of figure 3a. The peak at 665 nm disappeared after 180 min, see figure 3a. The MB degradation percentage values corresponding to x = 0.1, 0.3 and 0.5% samples were calculated taking into account the degradation of the peak at 665 nm and the equation (1). Figure 3b shows the MB degradation percentage produced by the Bi-doped samples as a function of time. The 0.1% Bi-doped sample was the most efficient photocatalyst, because it produced a degradation percentage of 96% after 240 min, followed by the 0.3at% sample with 91% of degradation for the same time. The 0.0% sample reached a maximum degradation of 84%. The blank solution presented 66% of degradation after 240 min. The sample 0.5at% Bi-doped sample showed a degradation of 76% [10]. We tried to make TiO_2: Bi but we obtained TiO/TiO_2:Bi instead. However, we obtained degradation results similar to these reported in reference [8], where the authors employed a complex method of sol-gel to make the TiO_2:Bi Powders and degrading Rhodamine B in aqueous solutions.

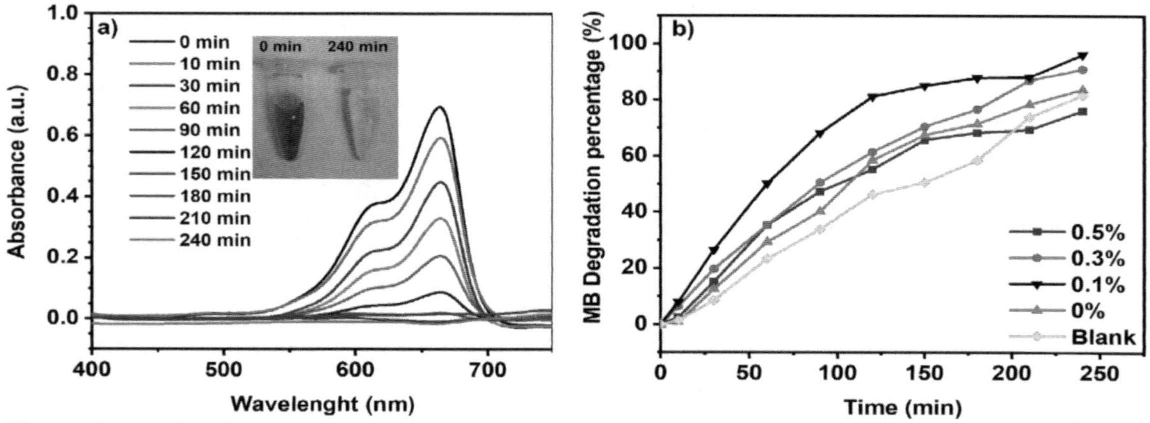

Figure 3. a) Absorbance spectra of the undoped TiO_2. b) Graph of MB degradation under sunlight irradiation.

4. Conclusions

The XRD patterns demonstrated that the crystalline structure of Bi doped titania changed with the content of Bi. If the Bi dopant is higher than 0.5at %, the composition changes from Bi doped TiO/TiO_2 to $Bi_{7.68}O_{12.16}Ti_{0.32}$ phase. The morphology presented in the undoped and Bi-doped samples (x=0.1 at%) showed coalesced grains and some porosity most similar to the typical morphology of the TiO_2. Samples doped with x=0.3 and 0.5 at% presented a more coalesced morphology. The highest MB degradation percentage was observed in the sample doped with 0.1 at%, which means that doping with bismuth is useful to enhance the degradation of MB. Due to the 91% of MB degradation in aqueous solutions, the TiO_2:$_x$Bi (x= 0.1at%) photocatalysts could be used in water-cleaning applications.

5. References

[1] S. Natarajan, H. C. Bajaj, and R. J. Tayade, "Recent advances based on the synergetic effect of adsorption for removal of dyes from waste water using photocatalytic process," *J. Environ. Sci.*, vol. 65, pp. 201–222, 2017.

[2] M. R. Al-mamun, S. Kader, M. S. Islam, and M. Z. H. Khan, "Photocatalytic activity improvement and application of UV-TiO_2 photocatalysis in textile wastewater treatment : A review," vol. 7, no. July, 2019.

[3] D. Zhu and Q. Zhou, "Action and mechanism of semiconductor photocatalysis on degradation of organic pollutants in water treatment : A review," *Environ. Nanotechnology, Monit. Manag.*, vol. 12, no. July, p. 100255, 2019.

[4] K. Wetchakun, N. Wetchakun, and S. Sakulsermsuk, "An overview of solar / visible light-driven heterogeneous photocatalysis for water purification : TiO_2 - and ZnO-based photocatalysts used in suspension photoreactors," *J. Ind. Eng. Chem.*, vol. 71, pp. 19–49, 2019.

[5] M. A. Fox and M. T. Dulay, "Heterogeneous Photocatalysis," *Chem. Rev.*, pp. 341–357, 1993.

[6] B. Yu, F. Meng, M. Wasim, R. Qin, and X. Liu, "Facile synthesis of AgNPs modi fi ed TiO_2 @ g-C_3N_4 heterojunction composites with enhanced photocatalytic activity under simulated sunlight," *Mater. Res. Bull.*, vol. 121, no. July 2019, 2020.

[7] T. N. Q. Trang, L. T. N. Tu, T. V Man, M. Mathesh, N. D. Nam, and V. T. H. Thu, "A high-efficiency photoelectrochemistry of Cu_2O / TiO_2 nanotubes based composite for hydrogen evolution under sunlight," *Compos. Part B*, vol. 174, no. May, 2019.

[8] H. Mao, Z. Jin, F. Zhang, H. He, J. Chen, and Y. Qian, "A high efficiency photocatalyst based on porous Bi-doped TiO_2 composites," vol. 44, no. May, pp. 17535–17538, 2018.

[9] W. Hao, F. Teng, Z. Liu, Z. Yang, Z. Liu, and W. Gu, "Synergistic effect of Mott-Schottky junction with oxygen defect and dramatically improved charge transfer and separation of Bi @ Bi-doped TiO_2," *Sol. Energy Mater. Sol. Cells*, vol. 203, no. December 2018, p. 110198, 2019.
[10] J. Ducke, M. Trömel, D. Hohlwein, and J. W. Kizler, "Yttrium and Titanium Bismuthates with Structures Related to Beta-$B_{i2}0_3$," *Acta Crystallogr.*, pp. 1994–1996, 1996.

Acknowldegements
D. Chavez acknowledges to CONACYT-Mexico for the Doctorate scholarship 693166. C.R. Garcia acknowledges to Universidad Autónoma de Coahuila for support this research. Author´s appreciate the technical services of Miguel Angel Savalza and Cristofer Álvarez Ramírez.

IOP Conf. Series: Materials Science and Engineering **842** (2020) 012015 doi:10.1088/1757-899X/842/1/012015

Potential Materials for Enhanced Oil Production from Northern oilfield, Thailand

K Maneeintr*, J Jiravivitpanya, T Meekoch and T Boonpramote

Carbon Capture, Storage and Utilization Research Group, Department of Mining and Petroleum Engineering, Faculty of Engineering, Chulalongkorn University, Bangkok 10330, Thailand

*E-mail: Krengkrai.M@chula.ac.th

Abstract. One of enhanced oil recovery techniques to produce more oil production is surfactant flooding. The surfactants are considered as the effective chemicals used in oilfield in Thailand. It is used to lower the interfacial tension (IFT) of two fluids and to increase mobility in the reservoir. For this study, various chemicals such as amine, fatty alcohol and glycerine are applied as the surfactants to reduce IFT between oil and water. Therefore, the aim of this work is to investigate the potential chemicals and measure the IFT reduction based on the conditions of subsurface at the oilfield in Thailand. These parameters such as temperature, pressure as well as the concentration of surfactant are adjusted to investigate the effects on IFT reduction. From the results, it is reported that pressure from 1,000 to 2,000 psi and temperature varied from 70 °C to 90 °C can reduce IFT insignificantly. However, types of chemicals and surfactant concentration are the main parameters that impact on the IFT reduction. Monoethanolamine at 4.0 wt. % can greatly decrease IFT up to 87 % for surfactant concentration. Finally, the results can be applied to use in the real field for enhanced oil production in Thailand.

1. Introduction

Currently, the energy demand is growing rapidly. Oil demand and production are required to serve this increasing demand. The oil production technology to satisfy this is the enhanced oil recovery (EOR) technology which is the tertiary recovery. It becomes more important because of the unpractical primary and secondary recoveries [1]. Chemical enhanced oil recovery is one of these kinds of technology. The surfactants are referred to as one of the practical flooding agents since it can remarkably lower the interfacial tension (IFT) between oil and water interfaces and enhance the performance of both displacement and sweep efficiencies. It can also lower the saturation of residual oil [2].

Nevertheless, some concerns still exist because they hinder the performance of flooding surfactants [3]. The effective in-situ surface active agent on chemical flooding is affected by several parameters such as reservoirs and operating conditions. It can cause the difficulty of how IFT changes under these conditions. However, the previous study [4] has reported some factors that influence the IFT. For this work, the research aims to study the various potential materials as surfactants used in enhanced oil production to product more oil especially oil from the northern area in Thailand. Furthermore, it is also focusing on the evaluation of the effects of variables such as concentration of surfactant, temperature and pressure on the interaction between hydrocarbon and aqueous solution. These effects will be useful to understand the mechanism of IFT reduction.

The IFT reduction in aqueous solution has the mechanism in term of colloidal aggregates called micelles. A micelle is an aggregate of surfactant molecules dispersed in a liquid colloid. A typical micelle in aqueous solution forms an aggregate with the hydrophilic parts in contact with the hydrophobic parts.

When the surfactant concentration is higher and the surfactant concentration called critical micelle concentration (CMC) is reached, the surfactant starts aggregating into micelles [1]. The CMC is the surfactant concentration at and above which micelles are formed. It can be determined for surfactant solutions by measuring the surface tension at different concentrations. Below the CMC, the surface tension decreases with increasing surfactant concentration as the number of surfactants at the interface increases. At above CMC, the surfactants will be insignificant for IFT reduction because it cannot help in IFT reduction. The CMC values will be accomplished when significant micelles concentration is suitable proportion [5].
Based on the previous work, there are some effective chemicals applied for IFT reduction such as salt mixed with surfactant [6] and antifoam [7]. These chemicals show the high performance in IFT reduction to very low level. For this study, to overcome the IFT reduction problem, the various potential chemicals such as amine, fatty acid and glycerine are applied as the surfactants to reduce IFT between oil and water because they are cheap and available in the market. Finally, the contribution of this work is that the project life of oil production can be extended and economically viable and it is used as basic knowledge and data to cover for oil production with reservoir and operating conditions.

2. Experiment

2.1 Chemicals

Oil sample is acquired from the oilfield from the northern area in Thailand. The density of this crude oil is 0.8462 g/cm^3 at 80 °C. The number of carbon atoms in oil is varied from 6 to more than 35. Monoethanolamide (MEA) and diethanolamide (DEA) are purchased from Sigma-Aldrich. Also, fatty alcohol and refined glycerine are obtained from Sigma-Aldrich Thailand. The chemical structure of all materials for IFT reduction are presented in Figure 1. From the molecular structure, it is evident that all structures have 2 main sections, a hydrophilic part (hydroxyl group) which is the polar section, the other part is called hydrophobic part (hydrocarbon chain) that is the non-polar section. Sodium chloride purchased from Ajax, Thailand is used as synthetic brine. It is obtained from distilled water mixed with sodium chloride as a main component to get the brine composition close to the original brine from the oilfield at various concentrations based on the original brine in the oilfield.

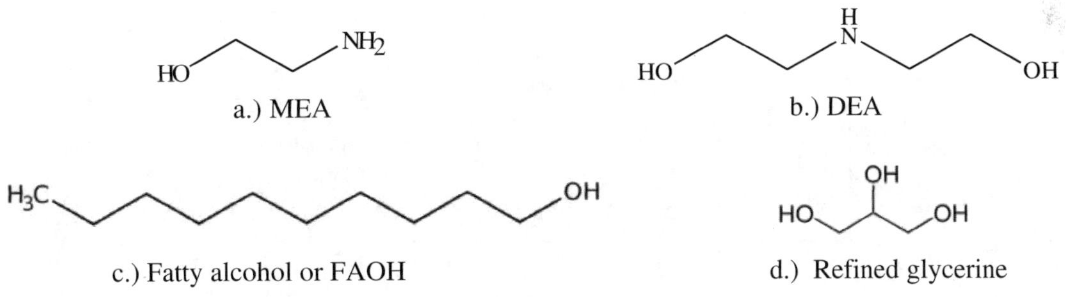

Figure 1. Chemical structure of materials for IFT reduction.

2.2 Equipment

The interfacial tension apparatus from Vinci Technology Company Model 700 is applied to investigate the interfacial tension (IFT) between the interface of liquid-liquid and liquid-gas systems in the chamber with the volume of 1×10^{-5} m^3. The maximum working conditions are 180 °C (or 350 °F) for temperature and 69 MPa (or 10,000 psi) for pressure. Moreover, a digital camera connected with a computer will take the picture and calculate the IFT. Also, the size of the drop is recorded and analyzed with Drop Analysis Software (DAS) supported by Vinci Technology.

2.3 Experimental procedure
The density of oil and the solution are measured at the beginning for software requirement by using density meter. The acetone and distilled water are used to clean the process of experiment. The solution and sample of oil are sent to the cylinders and temperature is set to the working condition. The surfactant solution and oil are fed to the chamber. The pressure gauge is applied to control to the designed pressure. The rising drop is formed and the camera linked to computer will take the pictures. The results will be recorded by the software provided by the manufacturer. The operating conditions for this work are shown in Table 1.

Table 1. IFT measurement operating conditions.

Parameters	Conditions
Surfactant type	MEA, DEA, Fatty alcohol, Refined glycerine
Concentration of solution (wt. %)	0.0, 0.5, 1.0, 2.0, 4.0
Temperature (°C)	70, 80 and 90
Pressure (MPa)	3.44, 6.88 and 10.32

3. Results and discussion

3.1 Effect of types of chemicals on IFT
Figure 2 presents the impact of the types of chemicals on IFT. It is evident that among these chemicals, MEA and DEA have the higher performance. It might be due to the size of the molecule are smaller and less complex compared to fatty alcohol and refined glycerine. Therefore, MEA and DEA can be potential candidates for IFT reduction and will be studied further in the real field in the future.

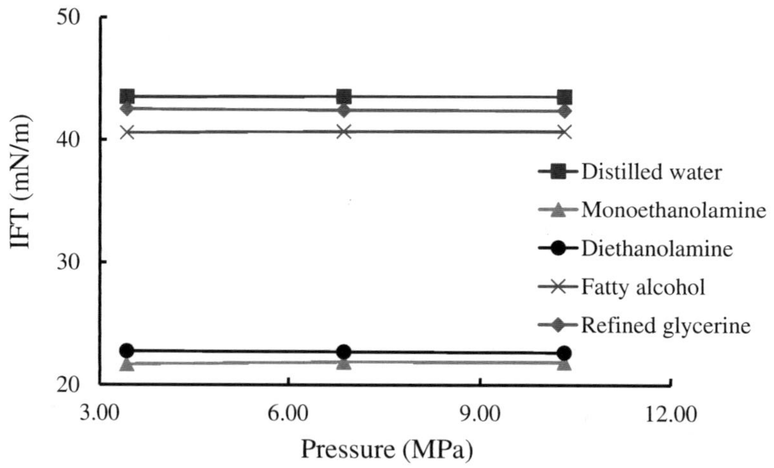

Figure 2. Effect of pressure on the IFT at concentration 0.5 wt. % and 80 °C.

3.2 Effect of pressure on IFT
Figure 2 shows the effect of pressure on IFT. It is clear that IFT has less impact as pressure increases. This is because this system is liquid-liquid system and pressure has less effect compared to gas phase system [8]. Therefore, IFT reduction cannot be seen clearly when pressure changes.

3.3 Effect of temperature on IFT
Figure 3 shows the effect of temperature ranging from 70 °C to 90 °C covering the temperature range in the oilfield. From Figure 3, IFT reduction is oppositely proportional to the temperature because at higher temperature, the free energy between solution and oil becomes lower thus making the mobility of oil and water enhanced and lowering the IFT [9]. Another reason is that the increase in temperature

can affect solution of surfactant due to the weakness of intermolecular forces at the interface [10]. MEA is the highest performance among other chemicals. The percent change of IFT are 9.18 % at temperature from 70 °C to 80 °C and 5.57 % at temperature from 80 °C to 90 °C at 1 wt.% MEA.

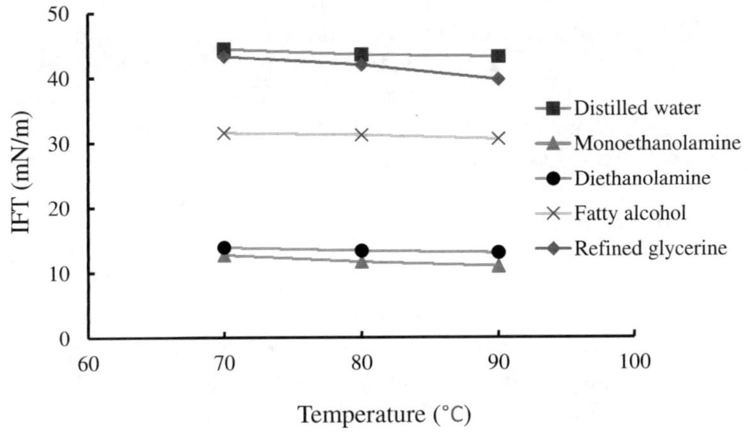

Figure 3. Effect of temperature on the IFT at 1 wt. % and pressure at 1,500 psi.

3.4 Effect of concentration on IFT

The effect of surfactant concentration is presented in Figure 4 with the concentrations ranging from 0 to 4 wt. % and 1,500 psi and 80 °C. The result illustrates that the IFT reduction comes from all cases because of each chemical characteristics. Based on molecular structure, there are two parts in the structure. One part is hydrophilic part (hydroxyl group) which can dissolve in water phase, and the other part is hydrophobic part (hydrocarbon chain) that can dissolve in oil-based phase. According to Prosser and Franses [11], the surface is covered by the surfactants molecules and the surface free energy (surface tension) becomes lower. Consequently, the higher surfactant concentration can dissolve more both in oil section and water section resulting in the IFT reduction.

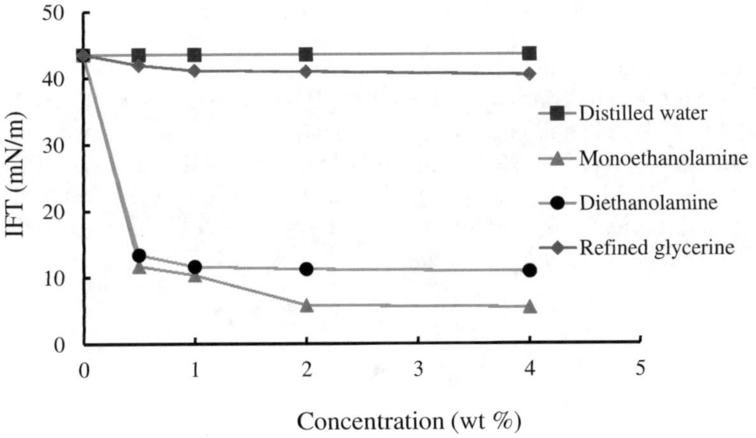

Figure 4. Effect of concentration of chemicals on the IFT at 1,500 psi and 80 °C.

In addition, MEA can reduce IFT at the lowest level and it is considered as the highest performance among this group of surfactants. Moreover, fatty alcohol cannot perform well at the concentration higher than 1 wt. % because it has the CMC effect on IFT reduction and it is not presented in this figure. The variation in concentration from 0%wt. to 4.0 wt. % can significantly decrease the IFT more than 87 % for MEA. Nevertheless, when the concentration is higher than 2.0 wt. %, the IFT value is relatively stable. This phenomena can be explained that the too much surfactant is put in a solution, thus; making the surfactant aggregating into micelles to the level above CMC and the amount of MEA is difficult to reduce IFT at the oil-water interface.

IOP Conf. Series: Materials Science and Engineering **842** (2020) 012015 doi:10.1088/1757-899X/842/1/012015

4. Conclusions

The chemical enhanced oil recovery techniques are applied to increase the oil production by using surfactant flooding. The surfactants are used as the effective chemical agents in the oilfield in Thailand. It is aimed to lower the IFT of two fluids and to make them flow easily in the reservoir. In this work, Monoethanolamide (MEA), diethanolamide (DEA), fatty alcohol and refined glycerine are commonly proposed as potential surfactants to decrease IFT between oil and brine based on the reservoir conditions from Fang oilfield. From the results, it is presented that the IFT data are measured and the effects of parameters such as surfactant concentration, temperature and pressure are also investigated at various conditions. Pressure ranging from 1,000 psi to 2,000 psi has less significant effect on IFT reduction at other constant conditions. In addition, the effect of temperature is relatively less effect and the rate of declining IFT trends to lower with the increase in temperature. However, the IFT can be decreased greatly with the types of surfactants. For this study, MEA shows the highest performance compared to other chemicals because of smaller size and less complex in the molecular structure. Furthermore, the IFT can significantly lower up to 87 % when the surfactant concentration becomes higher ranging from 0 to 4.0 wt. %. However, it will be relatively constant at the concentration higher 2.0 wt. %. The effects of each parameters will be useful to understand and used as basic knowledge to apply for oil recovery with reservoir conditions in the Fang oilfield, Thailand.

5. Acknowledgement

The authors would like to acknowledge Chevron Thailand Exploration and Production Ltd., for financial support of this project. The oilfield from the northern Thailand is acknowledged for oil and brine samples.

6. References

[1] Sheng J J 2011 *Modern Chemical Enhanced Oil Recovery: Theory and Practice* (Burlington: Gulf Professional Publishing)
[2] Ahmadi M A, Arabsahebi Y, Shadizadeh S R, Behbahani S S 2014 *Fuel* **117** 749
[3] Zhang Y P, Sayegh S G and Huang S 2007 *Canadian International Petroleum Conference*, Calgary, Alberta, Canada, June 12-14)
[4] Buckley J and Fan T 2005 *the International Symposium of the Society of Core Analysts*, Toronto, Canada, August 21-25
[5] Bera A, Mandal A and Guha B B 2014 *J. Chem. Eng. Data* **59** 89
[6] Schramm L, Stasiuk N. and Marangoni D G 2003 Physical Chemistry 99 10
[7] Karambeigi M S, Asl A H and Nasiri M 2014 *Journal of Petroleum Science and Technology* **4** 56
[8] Green D W and Willhite G P 1998 *Enhanced Oil Recovery* (Richardson : SPE)
[9] Xu W 2005 Master Thesis, Louisiana State University and Agricultural and Mechanical College
[10] Karnanda W, Benzagouta M S, Al Quraishi A and Amro M M 2013 *Arab J. Geosci.* **6** 3535
[11] Prosser A J and Franses E I 2001 *Colloids and Surfaces A: Physicochemical and Engineering Aspects* **178** 1

IOP Conf. Series: Materials Science and Engineering **842** (2020) 012016 doi:10.1088/1757-899X/842/1/012016

An Insight on Gas Migration in Cement Slurry

Pankaj Munjal, Kong Kian Hau*, and Cheng Chuen Hon Arthur

Civil and Environmental Engineering, National University of Singapore (NUS), Singapore

* ceekkh@nus.edu.sg

Abstract. The fluid invasion basically gas in the cement column is a worldwide problem. This is one of the main challenges of oil well industries to protect the wellbore from fluid invasion. This problem has recognized in the mid of 1970 and basic technologies associated with it has not changed significantly. The leakage of gas can cause to the catastrophic events such as blow out and contamination of adjacent aquifer layers leading to environmental concerns. Although much effort has been put forth on gas migration issues, but no efficient solutions have been established to control it. The main reason for this is some lack of understanding the phase behavior of cement slurry. This may be the reason for that there is no standard available till now for gas migration. This study reviews the various causes of occurrence of gas migration in the cement column. Moreover, this study provides a comprehensive insight on gas migration control additives, which is generally be used in oil well industries to protect the fluid invasion in cement slurry.

1. Introduction

Primary cementing operation requires placing of cement in the annular region between the metal casing and rock formation. A major issue after completion of the cementing operation is the invasion of gas from the formation into the cemented annulus resulting in uncontrolled flow of gas or fluid through the wellbore. The problem was first highlighted to industry by Carter and Slagle [1] in 1970 where they discussed the inability of a cement column to effectively maintain full hydrostatic pressure. Since then, several attempts have been made to study the gas flow in cements to achieve effective zonal isolation. Once the cement slurry is placed in the annulus, several processes begin to take place. The slurry begins to dehydrate due to a reduction in the liquid phase resulting in the formation of a gel phase. The gelation period is followed by setting process where cement hydration products begin to get packed in the cement matrix. During this process, the slurry gradually loses its ability to transmit hydrostatic pressure [2]. The pore pressure decreases and may lead to a situation where gas from the surrounding rock formation invades the cemented annulus. There are some other key properties of oil well cement (i.e., fluid loss, cement shrinkage, gel strength development, permeability, free-fluid, mechanical and chemical failure of cement sheath) which affect the invasion of gas migration.

Some solutions for gas migration control have been proposed. Chritsian et al. [3] proposed using cement slurries with good fluid loss control property in order to mitigate gas migration. This may be achieved by the use of special fluid loss control additives that hold the liquid phase within the cement matrix during setting process thereby maximizing the slurry's ability to transmit hydrostatic pressure. Griffin et al. [4] suggested the use of expansive cements that enable superior bonding and zonal isolation. Sabins et al. [5] indicated that the cement gelation period can be a crucial factor in preventing gas invasion into the annulus. A slurry with a shorter gel transition time changes from a fluid to a gel and finally to a solid state quick enough to prevent formation of gas channels in the cement matrix. In this study, causes of occurrence of gas migration is briefly described. Further, an insight on the gas migration additives used in oil well industries is presented.

2. Causes of gas migration

The one of the main reasons for invading the gas migration in the cement column is imbalance hydrostatic pressure [6]. The hydrostatic pressure falls below the formation pore pressure due to gel strength development, fluid loss, and cement shrinkage [7]. Several researchers [8-11] also reported another reason for gas migration in the cement column are: cement permeability, free fluid, mechanical and chemical failure of cement sheath. The schematic diagram of occurrence of gas migration in typical oil well is shown in figure 1.

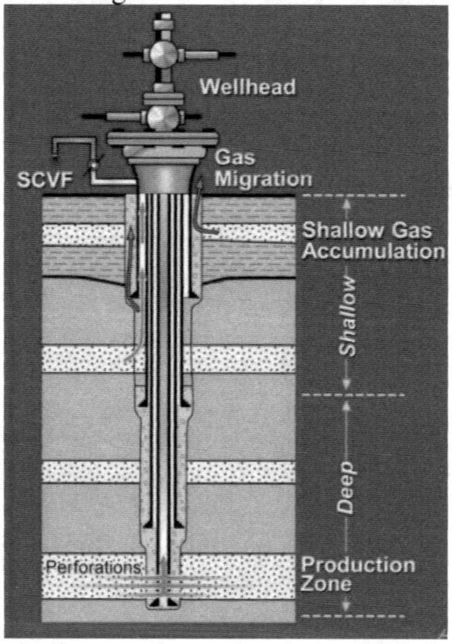

Figure 1. Schematic diagram of gas migration problem in typical oil well [12].

2.1. Gel strength development

The signal of development of gel strength start from the point at which the cement slurry begins to change from a true hydraulic fluid that transmit full hydrostatic pressure to a solid material that has measureable compressive strength [13]. Figure 2 shows the gel strength development of typical oil well cement slurry. As shown in figure 2, the slurry has sufficient fluidity in the fluid stage, and it maintains the higher hydrostatic pressure which is greater than the formation gas pressure. The hydrostatic pressure of cement slurry gradually decreases in the second stage (called gelled stage) with a simultaneous increase in static gel strength. The gas migration may occur if the hydrostatic pressure reduces below the formation gas pressure at point 'A' [14]. The static gel strength at point 'B' is generally about 50 Pa and when this value reach to about 240 Pa at point 'C', gas migration can be reasonably mitigated [14]. Once the second stage is complete the slurry starts to develop compressive strength at point 'E'. The time taken by cement slurry to reach point 'C' from the point 'B' is distinct as gel transition time. Thus, the potential of gases to invade in cemented annulus can be reduced by decreasing the gel transition time. However, the effect of gel transition time on gas invading is still not fully understood.

2.2. Fluid loss

Fluid loss is the leakage of liquid phase in the cement slurry under static (e.g. during tripping) and dynamic (when mud is circulating in the hole) conditions [15]. Excessive fluid loss decreases the hydrostatic pressure and contribute to provide the space in cement column for gas to enter [16]. American Petroleum Institute (API-10B) standard has described the limit of fluid loss (i.e., 50 mL / 30 min) which is often required to maintain adequate slurry performance. However, the fluid loss in the cement slurry can be reduced by fluid loss additives such as bentonite, water soluble polymers, cationic polymers, cellulose derivatives, carbonate powder, latex, asphaltenes, thermoplastic resin etc.

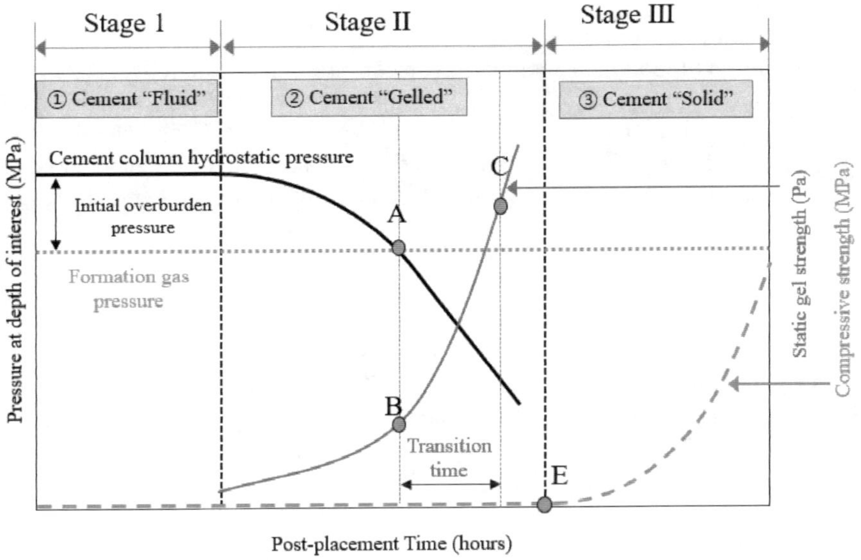

Figure 2. Gel strength development of typical oil well cement slurry [14].

2.3. *Chemical shrinkage*

The chemical shrinkage in oil well cement is divided into two types; (i) external shrinkage, (ii) total shrinkage. (i) The external shrinkage describes the bulk volume change of the cement slurry leading to a possible micro-annulus between the slurry and borehole wall; (ii) total shrinkage is the sum of contraction pores of the slurry and external shrinkage [17]. Numerous researchers [17-19] investigated the chemical shrinkage of oil well cement and they reported that the total shrinkage is varying from 0.6 to 6 Vol. % at 24 hours in the cement slurry. Backe et al., [17] observed that there is a close correlation between cement content and total chemical shrinkage. A low shrinkage and a short transition period can reduce the risk of gas migration. Pang et al. [19] reported that the total chemical shrinkage increases drastically with increasing the curing temperature, especially during early ages. Vu et al. [20] concluded that first chemical shrinkage occure then, swelling and autogenous shrinkage can occur, and their magnitude depend on degree of hydration, fineness of cement, w/c ratio and additive types.

2.4. *Cement permeability*

Permeability is the state of material which allow fluid (liquid or gas) to flow through it. The low permeability in the cement is one of the important criterion in the oil industries as permeability governs the flow of any external invading fluid [11]. Many researchers reported that the cement permeability is also a potential factor in gas migration incidents. However, cement permeability can be mitigated by use of additives. Sutton et al. [21] conducted various permeability test and found that the maximum cement permeability of 12 mD and that can substantially reduce by adding polymer fluid loss control additives. Goode [8] reported that permeability is not appeared to be a major factor at low temperatures because no measurable permeability has observed after certain period of time. Researchers [8, 21] emphasized that the other properties of cement particularly the fluid loss, gel strength and flow characteristics should be considered for gas migration properties.

2.5. *Free fluid*

Free fluid in the slurry pore can make the easy path for formation of gas or fluid. Many researchers [9-10] studied the effect of free water on the gas migration, giving inconsistent statements. Webster and Eikerts [9] observed that the free water in cement column has reduced the effective hydrostatic pressure which can make a path through the gas may migrate. On the contrary, Tinsley et al. [10] reported that the free water would not be a significant effect on invading of gas migration.

3. Gas migration theory

Several experimental and theoretical models have been presented to determine static gel strength and gas migration in cement slurries. Li et al. [22] attempted to predict gas migration potential based on the static gel strength and hydrostatic pressure reduction. Okoli [23] presented a model to evaluate slurry designs for effective zonal isolation and gas flow mitigation. These models utilize the static gel strength theory to describe gas flow potential at cement/formation or cement/metal interfaces. The hydrostatic pressure is dependent on the gelation tendency of the cement slurry and changes in hydrostatic pressure affects the ability of the cement column to restrict gas flow. Vazquez et al. [24] discussed a methodology in which the main parameters in evaluating gas migration are Flow Potential Factor (FPF) and Static Gel Strength (SGS). FPF is defined as [21]:

$$FPF = MPR/OBP \tag{1}$$

where, MPR: Maximum pressure restriction (psi), OBP: Hydrostatic overbalance pressure (psi)
The MPR is based on the pressure restriction due to SGS. For a general case, the pressure restriction due to SGS is defined as [21]:

$$P = (SGS \times L) / (300 \times D) \tag{2}$$

where, SGS: Static Gel Strength (lbf/100ft2), L: length of cement column (ft), D: Difference between open hole diameter and casing diameter (in).
A SGS value of 500 lbf/100ft2 is usually considered to be the strength sufficient to prevent fluid invasion through the cement column. In this situation, the pressure restriction due to SGS is at its maximum and equation (2) becomes [21]:

$$MPR = 1.67 \times L / (dhole\text{-}dcasing) \tag{3}$$

where, dhole: diameter of open hole (in), dcasing: diameter of metal casing (in).
Therefore, the FPF ratio as described by equation (1) can be used to evaluate fluid migration potential and the severity of gas flow potential can be determined from table 1 [25]:

Table 1. Classification of flow potential factor.

FPF	Severity Rating
< 4	Minor flow
$4 - 8$	Moderate flow
> 8	Severe flow

4. Gas migration control additives

This is the area in oil well industry that required more attentions as there are no specific details are available for use of gas migration control (GMC) additives. However, API STD 65 [26] has briefly described the need of GMC additives but specific details are not given. Researchers have used the following additives for gas migration.

4.1. Bentonite

Bentonite is a clay generated material frequently used in oil and gas industry due to its high swelling properties and low permeability for various applications. Few researchers have used the bentonite to mitigate the gas migration in the cement slurry. Ramadan et al. [27] have used the bentonite with latex additives to prevent the gas migration and reduce the fluid loss in cement slurry and observed that the gas transit time is decreased by 62 % with an increase in temperature.

4.2. Silica fume

Silica Fume is the one of the common gas migration additives. A very fine (grey colored) non-crystalline silica produced in electric arc furnace as a by-product of the production of elemental silicon metal or ferrosilicon alloys. Silica fume particles are very small, with maximum particles size less than

1 μm (100 times finer than cement particle) with high specific surface area 20 m2/g [28]. Various researchers [29-30] have studied the application of silica fume for mitigation of gas migration and concluded that silica fume is very effective for gas migration. Skalle and Svenn [29] reported that micro-silica is effective for gas migration for temperature up to 150°C. Daou and Piot [30] observed that only moderately compacted micro-silica with a bulk density of approximately 300 kg/m3 would be helpful in developing a good cement performance.

4.3. Latex

Latexes are aqueous dispersions of polymer particles (such as surfactants) which imparts stability to the dispersion [31]. Latex is most commonly additives used for gas migration. Most latexes, however, can loose their elastic ability and become brittle at higher temperatures (> 160°C) [29]. Latex fills the pores in OWC slurry and decrease the permeability of the cement. The most common latex are vinylidene chloride, poly-vinylacetate, and styrene-butadiene,

4.4. Carbon black

Carbon black is a product of incomplete combustion of coal tar or ethylene cracking tar in the form of paracrystalline carbon which can be used as gas migration additives. The range of the particle size of the carbon black is 10 to 200 nM. Calloni et al. [32] reported that carbon black (4 % by weight of cement) is sufficient to control gas migration in all formulation.

4.5. Hydroxypropyl methylcellulose (HPMC)

HPMC is a cellulose based gelling agent which can be used for gas migration agent. Abbas et al. [33] reported that HPMC can create an impervious barrier for water and prevent the gas migration in the wellbore hole.

5. Conclusions

A comprehensive review on causes of occurrence of gas migration and additives used in oil well industries to control this problem in the wellbore is discussed. The following conclusion is drawn based on review.

1. The invasion of gas can be prevented by keep the cement pore pressure more than the formation gas pressure.
2. The phenomenon of gas migration is believed to occur during the transition state between initial and final set of the cement.
3. The invasion of gas into cement matrix can be prevented by use of suitable polymeric materials and additives in the cement slurry to immobilize the fluid within the pore spaces.
4. The additives such as bentonite, silica fume, and latex are generally used in oil industries to control the gas migration.
5. More research work is required to fully understand the phenomenon behind the occurrence of gas migration in cement column and its solutions.

6. References

[1] L. G. Carter, and K. A. Slagle, *SPE-3164-PA,* 1170-74 (1972).
[2] P. R. Cheung and R. M. Beirute, *J. Pet. Tech.*, ***37 (6),*** 1041-1048 (1985).
[3] W. W. Christian, J. Chatterji, and G. W. Ostroot, *J. Pet. Tech.,* **28 (11)**, 1361-69 (1976).
[4] T. J. Griffin, L. B. Spangle, and E. B. Nelson, *Oil and Gas J.,* 143-51 (1979).
[5] F. L. Sabins, J. M. Tinsley, and D. L. Sutton, "Transition time of cement slurries between the fluid and set state," SPE technical conference and exhibition, Dallas, US (1980).
[6] P. M. Mohammad and J. Moghadasi, *SPE-105663-MS*, 1-8 (2007).
[7] H. Justnes, D. Van Loo, B. Reyniers, P. Skalle, and J. Sveen, *Advan. Cem. Res.,* **7 (26)**, 85–90 (1995).
[8] J. M. Goode, *J. Pet. Tech.* **14 (08)**, 851–54 (1962).
[9] W. W. Webster and J. V. Eikerts, *SPE-8259-MS,* 1-8 (1979).
[10] J. M. Tinsley, E. C. Miller, F. L. Sabins, and D. L. Sutton, *J. Pet. Tech.,* **32 (8)**, 1427–37 (1980).

[11] S. Appleby and A. Wilson, Chem. Eng. Sci., **51(2)**, 251-267 (1996).
[12] K. Muehlenbachs, http://www.frackingcanada.ca/industrys-gas-migration/ (Accessed 9 November 2019).
[13] J. Moon, and S. Wang, *SPE 55650-MS*, 1-10, (1999).
[14] N. Lee, A. Prabhakar, O. K. C. Gary, J. Moon, M. H. Zhang, C. H. A. Cheng, H. Klaus, and K. K. Hau, Const. Build. Mater., **191**, 1093-1102 (2018).
[15] O. John, *Amer. J. Eng. Res.*, **6 (8)**, 136-151 (2017).
[16] Z. Li, J. Vandenbossche, A. Iannacchione, J. Brigham, and B. Kutchko, *SPE Drilling & Comp.*, **31(2),** 145–58 (2016).
[17] K. R. Backe, P. Skalle, O. B. Lile, and S. K. Lyomov, *J. Can. Pet. Tech.,* **37** (1998).
[18] M. E. Chenevert and B. K. Shrestha, *SPE Drilling Eng.*, **6**, 37-43 (1991).
[19] X. Pang, C. Meyer, G. P. Funkhouser, and R. Darbe, *Const. Build. Mater.,* **74**, 93–101 (2015).
[20] M. H. Vu, A. P. Bois, and A. Badalamenti, *SPE-189384-MS*, 1-24, (2018).
[21] D. L. Sutton, F. Sabins, and R. Faul, *Oil Gas J.,* **82**,1984.
[22] Z. Li, J. Vandenbossche, A. Iannacchione, J. Brigham, and B. Kutchko, *SPE Drilling & Comp.*, **31**, 1-14 (2016)
[23] U. Okoli, *SPE-193490-MS*, 1-12 (2018).
[24] G. Vazquez, A. M. Blanco, and A. M. Colina, *SPE-94901-MS*, 1-9 (2005).
[25] E. Nelson, "Well Cementing," Schlumberger Dowell, Sugar Land, Texas, 1990.
[26] API STD 65- 2, *Standard for isolating potential flow zones during well construction*, American Petroleum Institute, (2010).
[27] M. A. Ramadan, S. Salehi, G. Kwatia, C. Ezeakacha, and C. Teodoriu, *J. Pet. Sci. Eng.,* **179**, 26-35 (2019).
[28] E. Grabowski and E Gillott, *Cem. Concr. Res,.* **19**, 333-344(1989).
[29] P. Skalle and J. Sveen, *SPE-23075-MS,* 1-12 (1991).
[30] F. Daou and B. M. Piot, *SPE-112701-PA,* **24**, 1-9 (2009).
[31] G. Kwatia, C. Ezeakacha, and S. Salehi, "Literature report of elastomer sealing materials and cement systems," BSEE Project No.: E17PC00005, 1-78 (2017).
[32] G. Calloni, N. Moroni, and F. Miano, *SPE-28959-MS*, 1-9 (1995).
[33] G. Abbas, S. Irawan, S. Kumar, M. N. Khan, and S. Memon, *SPE-169643-MS, 1-10 (*2013).

Funding and acknowledgments
This work was supported by the Singapore Maritime Institute (ID: SMI-2016-OF-05). Grateful acknowledgement is made to Nautec for collaborative support with access to the company's OWC laboratory, equipment and materials. The authors would like to acknowledge key personnel from Nautec: Dr Kok-Seng Chia, and Mr Elo Yde, for their advisory role in this project.

An Experimental Study on the Technique Optimization in Localization Process of 4J36 Invar Alloy-A Micro-Electrochemical Machining Perspective

Jianxiao Bian[1,2,*], Baoji Ma[2], Yanwei Hu[2], Zhichen Jing[1], Xin Liu[1]

[1]School of Mechanical Engineering ,Longdong University , Qingyang 745000, China
[2]School of Mechanical Engineering ,Xi'an Technological University , Xi'an 710021, China
E-mail:Jxbian@ldxy.edu.cn

Abstract. This study introduces electrochemical processing into the processing of 4J36 Invar alloy to obtain higher processing accuracy. An experimental study based on central composite design (CCD) was carried out to analyze the effects of voltage, duty cycle, feed rate, and inter-electrode gap on the localization of Invar processing. The response surface method is used to establish a mathematical model between process parameters and processing locality. The model shows that the minimum Side Gap is 158.7μm, and the prediction results are verified by experiments. The prediction error of the mathematical model is 1.44%, which means that the proposed optimal mathematical model can solve the optimal processing parameters of 4J36 Invar alloy.It has a positive significance for the wide application of Invar.

1. Introduction

Electrochemical processing, also known as Electrochemical Machining (ECM).is a process in which an anodic dissolution of a metal workpiece occurs in an electrolyte[1,2]. The whole process is completed by the synergistic action of electrochemical, flow field and electric field. It has the advantages of high productivity, no tool loss, no cutting stress, and the principle of ion erosion makes it have the potential of fine and precise processing. Electrochemical machining is suitable for the processing of difficult-to-machine materials and complex shapes. It has been widely used in the manufacturing industries of aviation, aerospace and molds[3]. Invar is an alloy containing 64% iron and 36% nickel. Its coefficient of expansion is very low, with a value of 1.8x10-6cm/°C and a temperature ranging from a low temperature of -196°C to a temperature of 260°C.This value is twice as low as stainless steel and 100 times lower than iron, which makes Invar alloy widely used in various fields[4,5]. The process parameters in electrochemical machining have an important influence on the processing precision of Invar. S.H.Kim et al.[6]The surface quality of Invar were investigated under different voltage and electrode shapes. Choi S-G et al.[7]The current density in electrochemical machining was studied by finite element analysis. K.H.Chun et al.[8]The processing properties of different electrolyte compositions for electrochemical processing of Invar alloys were investigated. The electrolyte components under the optimum process conditions were determined. However, the influence of different process parameters on the accuracy of electrochemical processing of Invar alloy has paid little attention.

In this paper, the influence of electrochemical processing parameters on the localization of 4J36 Invar alloy is studied. The experimental study based on the central composite design (CCD) is carried out to analyze the effects of voltage, duty cycle, feed speed and gap between electrodes on the localization of 4J36 Invar alloy processing. In the experiment, the localization is represented by the processing Side

IOP Conf. Series: Materials Science and Engineering **842** (2020) 012017 doi:10.1088/1757-899X/842/1/012017

Gap, and the mathematical model between the processing parameters and the localization is established by the response surface method[9]. To solve the problem of optimum processing conditions.

2. Experiment

2.1. Experimental Set-Up

The structure of the ECM experimental system is shown in Fig. 1 and 2. The experimental system consists of four subsystems: the power supply system, the machine tool system, the microelectrode system, and the processing control and monitoring system. The power supply system can provide variable pulse voltage, duty cycle, pulse frequency. The machine tool system ensures high precision in micro electrochemical machining. The microelectrode system consists of a tool electrode, an electrolytic cell, an Invar alloy. The 4J36 Invar workpiece is fixed on the electrolyzer and placed on the lifting platform. The machining control and monitoring system has a motion control card and Supereyes. supereyes monitors the machining process and captures images. The material used in the experiment was 4J36 Invar. table 1 lists the chemical composition of 4J36 Invar.

Table 1. Chemical compositions of magnesium 4J36 invar alloy.

Components	Fe	Ni	Cr	C	Mn	Si	Co	P	S
Percents(%)	62.64	36	0.2	0.03	0.4	0.2	0.5	0.02	0.01

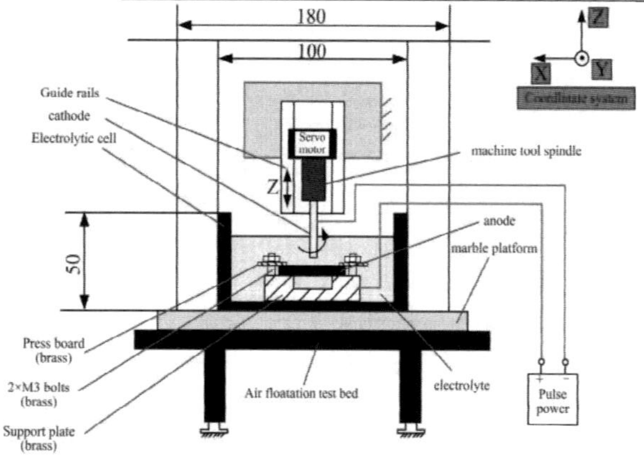

Figure 1. Experimental system of ECM.

Figure 2. Experimental equipment.

2.2. Experimental Arrangement

Voltage (6, 8, 10V), duty cycle (40, 60, 80%), feed rate (0.5, 1.5, 2.5μm/s) and inter-electrode gap (100, 150, 200μm) are used in electrochemical machining. As the main processing technology parameters, the test is designed based on the central composite design, considering three factors and three levels. Therefore, 30 sets of tests (including 6 sets of repeated tests) were carried out, and the microelectrode system is shown in Table 2. The specific experimental steps are:

a) The anode workpiece is placed in an ultrasonic cleaner for ultrasonic cleaning to remove surface impurities to prevent interference with the experiment.

b) Place the workpiece on the electrolyzer fixture and install the tool cathode to adjust the inter-electrode gap.

c) Pour the configured 1mol/L $NaNO_3$ + 0.1mol/L $C_6H_8O_7$ electrolyte into the electrolytic cell to soak the workpiece for 5-10 minutes. The workpiece and the electrolytic cell are cleaned after processing.

d) The Side Gap of the machined hole was observed using a scanning electron microscope and the results were recorded (Side Gap equals hole radius minus cathode radius).

2.3. Result and Discussion

The design of the test and the test results obtained are shown in Table 3. The experimental results show the influence of process parameters on the Side Gap, as shown in Fig 3. It can be seen that the Side Gap increases as the pulse voltage and duty cycle increase. It decreases as the feed rate increases. The inter-electrode gap has less influence on the Side Gap. The value of the Side Gap varies within a small range.

Table 2. Experiment parameters.

parameter	values
Size of cathode	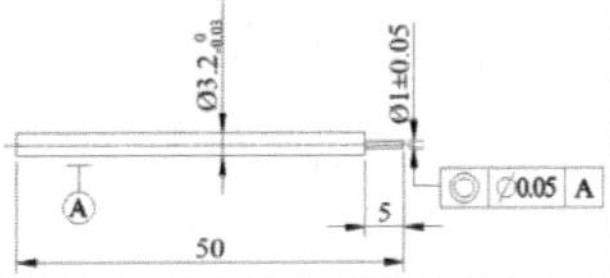
Size of anode	25mm×25mm×0.3mm
Cathode material	H62 brass
Anode material	4J36 invar alloy
Electrolyte	1mol/L $NaNO_3$+0.1mol/L $C_6H_8O_7$
Spindle speed	4000 rpm

3. Optimization

3.1. Development RSM models

The response surface method is used to establish a mathematical model of the actual factors between the process parameters and the machining Side Gap. The model includes linear correlations and interactions. As shown below:

$$Side.Gap = 182.63 + 9.56 \times A + 4.22 \times B - 11.44 \times C + 1.5 \times D + 3.44 \times AB + 2.31 \times AD - 2.94 \times BC \quad (1)$$

Among them, A, B, C, and D are voltage, duty ratio, feed speed, and inter-electrode gap, respectively. In order to determine the reliability of the model and compare the influence of process parameters on the Side Gap, a second-order model analysis of variance was performed. Table 4 shows the ANOVA results for the Side Gap. Among them, the R^2 term is 0.9213, which proves the reliability of the model.

3.2. Parametric influence

As shown in Fig 4. the effect of process parameters synergistically on the Side Gap can be seen.

3.3. Verification

According to the mathematical model, the processing parameters and predicted values of the minimum Side Gap can be obtained, and the results of experimental verification are shown in Table 5. It can be seen that the error is 1.44%.The fig 5. shows the processing results of the optimized parameters.

IOP Conf. Series: Materials Science and Engineering **842** (2020) 012017 doi:10.1088/1757-899X/842/1/012017

Table 3. Experiment result.

No	Voltage	Duty Cycle	Feed Speed	Interpolar Gap	Side Gap
1	10	80	2.5	200	194
2	8	40	1.5	150	178
3	8	60	0.5	150	193
4	8	60	1.5	150	183
5	10	80	0.5	100	209
6	6	40	0.5	200	182
7	10	40	2.5	200	175
8	8	60	1.5	150	180
9	6	40	0.5	100	181
10	8	60	1.5	100	175
11	10	60	1.5	150	192
12	10	80	2.5	100	181
13	8	60	1.5	150	186
14	6	80	0.5	200	189
15	8	60	2.5	150	174
16	10	80	0.5	200	218
17	8	60	1.5	150	185
18	8	60	1.5	200	175
19	6	80	2.5	200	152
20	8	60	1.5	150	179
21	6	80	2.5	100	164
22	6	40	2.5	100	160
23	10	40	0.5	200	201
24	8	80	1.5	150	189
25	6	40	2.5	200	169
26	10	40	2.5	100	178
27	8	60	1.5	150	182
28	6	60	1.5	150	175
29	10	40	0.5	100	188
30	6	80	0.5	100	192

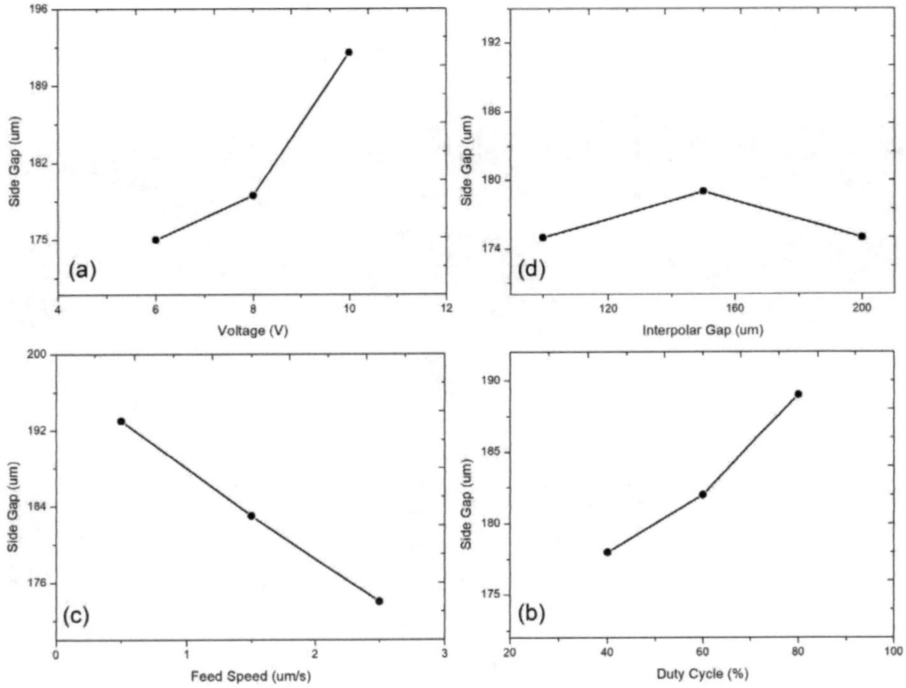

Figure 3. Effect of process parameters on Side Gap.

Table 4. ANOVA results of Side Gap.

Source	*Sum of Squares*	*df*	*Mean Square*	*F-value*	*p-value*	
Model	4775.19	7	682.17	36.80	< 0.0001	significant
A	1643.56	1	1643.56	88.67	< 0.0001	
B	320.89	1	320.89	17.31	0.0004	
C	2357.56	1	2357.56	127.19	< 0.0001	
D	40.50	1	40.50	2.19	0.1535	
AB	189.06	1	189.06	10.20	0.0042	
AD	85.56	1	85.56	4.62	0.0429	
BC	138.06	1	138.06	7.45	0.0122	
Residual	407.78	22	18.54			
Lack of Fit	370.28	17	21.78	2.90	0.1212	not significant
Pure Error	37.50	5	7.50			
Cor Total	5182.97	29			R^2	0.9213

Table 5. Results of confirmatory experiment.

Exp	*Parameters settings*				*Side Gap*		
No	A	B	C	D	Exp	Pred	Error%
1	6	80	2.5	200	161	158.7	1.44%

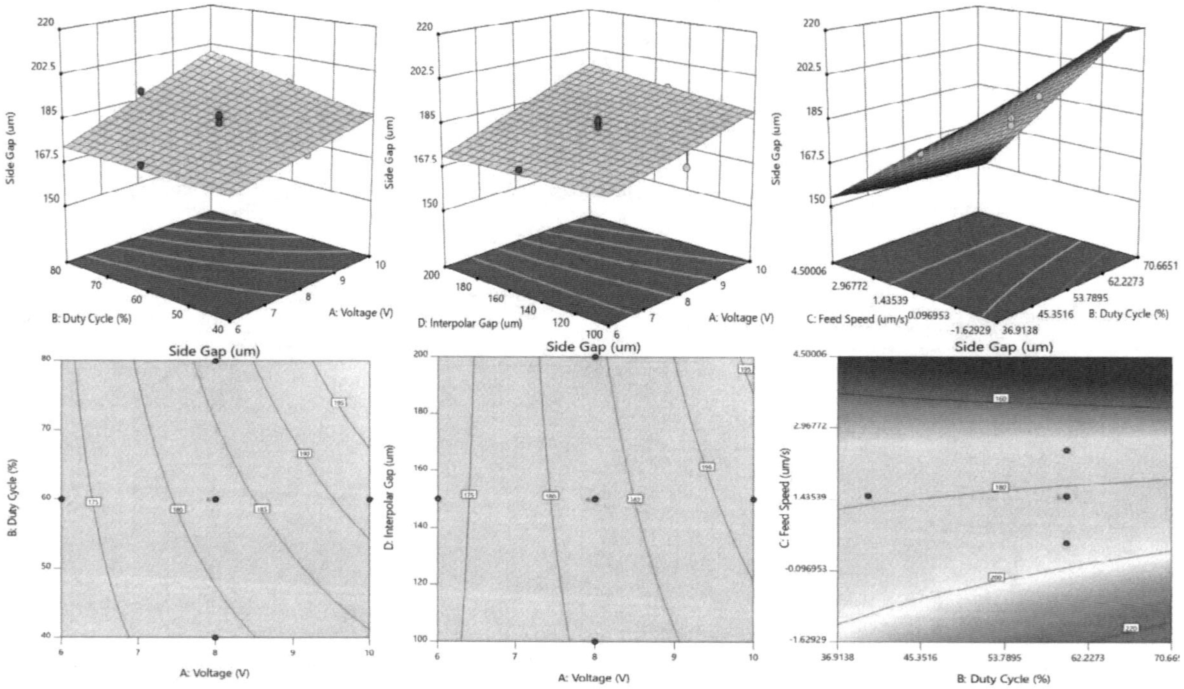

Fig.4. Response Chart of Synergistic Effect of Process Parameters on Side Gap.

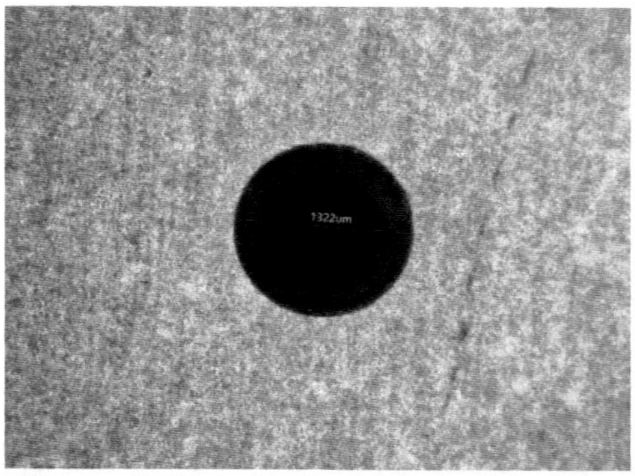

Figure 5. Optimized minimum Side Gap.

4. Conclusions

- The factor that has the greatest influence on the Side Gap of the hole is the feed speed, followed by the pulse voltage. The inter-electrode gap has less effect on the Side Gap.
- A response diagram of the synergistic effect of the process parameters on the Side Gap of the hole is obtained.
- Establish a mathematical model of process parameters and pore Side Gap in the process of micro-electrochemical machining of 4J36 Invar alloy. The optimum process parameters were determined and verified by experiments with an error value below 5%.

Acknowledgments

This study was supported by Doctoral Foundation Project of Longdong University(Grant No. XYBY1905).

Reference

[1] Unune D R, Mali H S. Current status and applications of hybrid micro-machining processes: A review[J]. Proceedings of the Institution of Mechanical Engineers, Part B: Journal of Engineering Manufacture, 2014, 229(10): 1681-1693.

[2] Rahman Z, Das A K, Chattopadhyaya S. Microhole drilling through electrochemical processes: A review[J]. Materials and Manufacturing Processes, 2017, 33(13): 1379-1405.

[3] Davydov A D, Kabanova T B, Volgin V M. Electrochemical machining of titanium. Review[J]. Russian Journal of Electrochemistry, 2017, 53(9): 941-965.

[4] Choi W-K, Kim S-H, Lee E-S. Electrochemical micro machining characteristics of Fe64Ni36 invar film using micro WC rod electrode[J]. Microsystem Technologies, 2016, 23(2): 405-410.

[5] Kim S H, Choi S G, Choi W K, et al. Surface characteristics of Invar alloy according to micro-pulse electrochemical machining[J]. Materiali in tehnologije, 2017, 51(5): 745-749.

[6] Kim S H, Choi S G, Choi W K, et al. Pulse electrochemical machining on Invar alloy: Optical microscopic/SEM and non-contact 3D measurement study of surface analyses[J]. Applied Surface Science, 2014, 314: 822-831.

[7] Choi S-G, Kim S-H, Choi W-K, et al. A study of the current density analysis for two type method on the invar alloy in electrochemical machining[J]. Journal of Mechanical Science and Technology, 2017, 31(9): 4345-4351.

[8] Chun K H, Kim S H, Lee E S. Analysis of the relationship between electrolyte characteristics and electrochemical machinability in PECM on invar (Fe-Ni)fine sheet[J]. The International Journal of Advanced Manufacturing Technology, 2016, 87(9-12): 3009-3017.

[9] Mehrvar A, Basti A, Jamali A. Optimization of electrochemical machining process parameters: Combining response surface methodology and differential evolution algorithm[J]. Proceedings of the Institution of Mechanical Engineers, Part E: Journal of Process Mechanical Engineering, 2016, 231(6): 1114-1126.

IOP Conf. Series: Materials Science and Engineering **842** (2020) 012018 doi:10.1088/1757-899X/842/1/012018

Injection Molding Study on the Strength of Weld Line for Polycarbonate and Carbon Fiber Composite Material

Chen-Yuan Chung[1,*], Chin-Chang Yeh[2], Chia-Min Lin[1]

[1] Department of Mechanical Engineering, National Central University, Taoyuan City, Taiwan
[2] Department of Mechanical Engineering, Oriental Institute of Technology, New Taipei City, Taiwan

*E-mail: cxc474@case.edu

Abstract. In this study the polycarbonate (PC) and carbon fiber (CF) are used as the experimental materials to investigate the impact of injection molding process parameters on tensile strength of weld line using Taguchi method and response surface methodology. The experiments are conducted via 27 sets of parameter combinations designed based on carbon fiber content, melt temperature, mold temperature, injection speed, injection pressure, packing speed, and packing pressure. The process parameter combination with the highest tensile strength is identified using Taguchi method. Later on, the response surface methodology is used to establish the tensile strength predicting formula formed by process parameters in order to understand the correlation between process parameters and tensile strength. The experimental and analysis results indicate the melt temperature and mold temperature have the most significant impacts on tensile strength.

1. Introduction

During the injection molding process, the forming of weld line is generated by the merging of two flow fronts of plastic melts. In the molecular bonding region, the directions of molecular chains of melts are in parallel with each other, thus reducing the molecular bonding in the weld region and resulting in poorer mechanical property [1] in the weld region. Therefore, there are many scholars hoping to predict the structural strength of weld line via derivation of formula. The theoretical molecular chain self-diffusion model proposed by Kim and Suh [2] in 1986 has been wildly applied to the calculation of prediction of mechanical strength and birefringence of weld line. In addition to the analysis of weld-line strength via theoretical model, there were also many scholars investigating the impacts of manufacturing processes on weld-line strength. Chen et al [3] investigated the impacts of different parameters, such as melt temperature, mold temperature, injection speed, and packing pressure, on weld-line strength of ABS material. It was found that weld-line specimens molded at higher melt temperature, higher mold temperature, faster injection speed, and lower packing pressure would lead to better mechanical properties. In recent years, U.S. Department of Energy has been applying the manufacturing of long-carbon-fiber reinforced thermoplastic composites to lightweight materials [4]. Yan and Cao [5] investigated the structure and interfacial shear strength of glass fiber/carbon fiber reinforced polypropylene hybrid composites fabricated by the direct fiber feeding injection molding process. The process could improve the fiber lengths and orientation distributions and thus increasing the shear strength.

This study presents an experimental approach followed by rigorous statistical techniques to identify the optimal combination of process parameters that can maximize the weld-line strength of PC composite reinforced with CF. The Taguchi method and response surface methodology are used to investigate the impacts of injection molding process parameters on the weld-line strength of CF/PC composite. Additionally, the tensile strength of weld line is verified by the experiments conducted based on optimal injection molding parameters. The weld line is vulnerable to failure due to reduced mechanical strength. It is expected that the manufacturing process optimization of this paper can serve as the reference for quality improvement of plastic product.

2. Materials and Methods

2.1. Experimental equipment

The pure PC and CF/PC used in this study are produced by Kotec Corporation (Osaka, Japan). First the hopper dryer is used for four hours of dehumidifying at the high temperature of 120°C, and then the injection molding machine (SM150HCV, Asian Plastic Machinery Co., Ltd.) is used in coordination with mold temperature controller (TOD-200, Wedlon Automation Co. Ltd.) to produce test specimen. Finally, the tensile tests were performed on a universal testing machine (AI-7000M, Gotech Testing Machines Inc.) at a crosshead speed of 5 mm/min according to ASTM D638 (Type I) standard. Furthermore, analysis of experimental data was performed based on Taguchi method and response surface methodology using Minitab software (Minitab Inc., State College, PA, USA).

2.2. Taguchi method

The Taguchi method was applied to determine the effects of process parameters on the weld-line strength and the optimal set of parameters that would maximize the weld-line strength. An L_{27} (3^7) orthogonal array was selected for the experimental design for each of the seven factors. The three levels for the seven parameters were identified during the 27 experiments (Table 1). In order to maximize the weld-line strength, the formula of S/N ratio describing the larger-the-better characteristic was based on the following equation.

$$S/N = -10\log\left(\frac{1}{n}\sum_{i=1}^{n}\frac{1}{y_i^2}\right) \tag{1}$$

where y_i is the measured property for the *i*th tensile test, n is the number of tests. The analysis of variance (ANOVA) was performed using Minitab to obtain the effects of the process parameters on mechanical property. In this way, optimal levels of the process parameters can be estimated.

2.3. Response surface methodology

Different from the Taguchi method, response surface methodology basically combines statistical and experimental methods with data-fitting techniques. Based on the response acquired in the experiments, regression analysis is utilized to identify the relationships between the response and the design variables. The mathematical relationship of the Y response on the corresponding parameters is usually expressed by the following second-order polynomial equation [6, 7]:

$$Y = \beta_0 + \sum_{i=1}^{k}\beta_i x_i + \sum_{i=1}^{k}\beta_{ii}x_i^2 + \sum_{i<j}\sum \beta_{ij}x_i x_j + \varepsilon \tag{2}$$

where Y is the observed response (mechanical property), k is the number of design variables, β_0 is a constant. β_i, β_{ii}, β_{ij} are the linear, quadratic and interaction coefficients, respectively. x_i and x_j are the design variables that correspond to the injection process parameters. ε is the random error.

Table 1. Orthogonal array $L_{27}(3^7)$ of the experimental runs for Taguchi method.

run	Factor A Carbon fiber (%)	Factor B Melt temperature (℃)	Factor C Mold temperature (℃)	Factor D Injection pressure (bar)	Factor E Packing pressure (bar)	Factor F Injection speed (mm/s)	Factor G Packing speed (mm/s)
1	0	270	50	50	10	20	20
2	0	270	50	50	30	30	30
3	0	270	50	50	50	40	40
4	0	290	70	70	10	20	20
5	0	290	70	70	30	30	30
6	0	290	70	70	50	40	40
7	0	310	90	90	10	20	20
8	0	310	90	90	30	30	30
9	0	310	90	90	50	40	40
10	5	270	70	90	10	30	40
11	5	270	70	90	30	40	20
12	5	270	70	90	50	20	30
13	5	290	90	50	10	30	40
14	5	290	90	50	30	40	20
15	5	290	90	50	50	20	30
16	5	310	50	70	10	30	40
17	5	310	50	70	30	40	20
18	5	310	50	70	50	20	30
19	10	270	90	70	10	40	30
20	10	270	90	70	30	20	40
21	10	270	90	70	50	30	20
22	10	290	50	90	10	40	30
23	10	290	50	90	30	20	40
24	10	290	50	90	50	30	20
25	10	310	70	50	10	40	30
26	10	310	70	50	30	20	40
27	10	310	70	50	50	30	20

3. Results and Discussions

The tensile specimen of injection molding is fabricated according to the regulation of ASTM-D638 TYPE I, and the injection molding experiment is conducted according to $L_{27}(3^7)$ orthogonal array. Five

IOP Conf. Series: Materials Science and Engineering **842** (2020) 012018 doi:10.1088/1757-899X/842/1/012018

samples are molded under each set of molding conditions, and such sample is the tensile specimen fabricated according to the ASTM-D638 Standard (Figure 1).

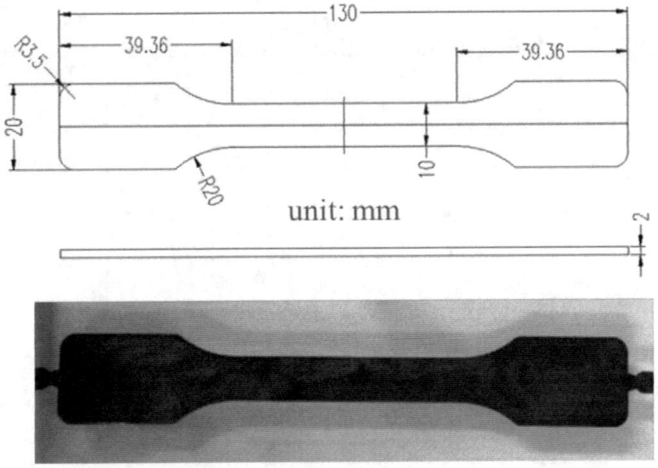

Figure 1. Geometry and dimension of tensile test specimen.

At the beginning the Taguchi method is used, and the L_{27} (3^7) orthogonal array is used to arrange different process parameters for investigating the tensile strength of test specimen (Table 2), and then the optimal process parameter combinations can be obtained via the S/N response graphs of tensile strength against various factors (Figure 2) in order to enhance the mechanical property of finished product. The factor response graph (Figure 2) reveals that the best tensile strength process parameter combination of weld-line specimen is "A3-B3-C2-D1-E1-F2-G1". The analysis of variance (ANOVA) is used to evaluate the experimental error and significance test of factor. The ranking of influences of various process parameters on tensile strength is as shown in Table 3. In the end, the experiment is conducted via the optimal process parameter combination obtained by Taguchi method, "A3-B3-C2-D1-E1-F2-G1", and it is proven that the tensile strength is as high as 82MPa (Figure 3).

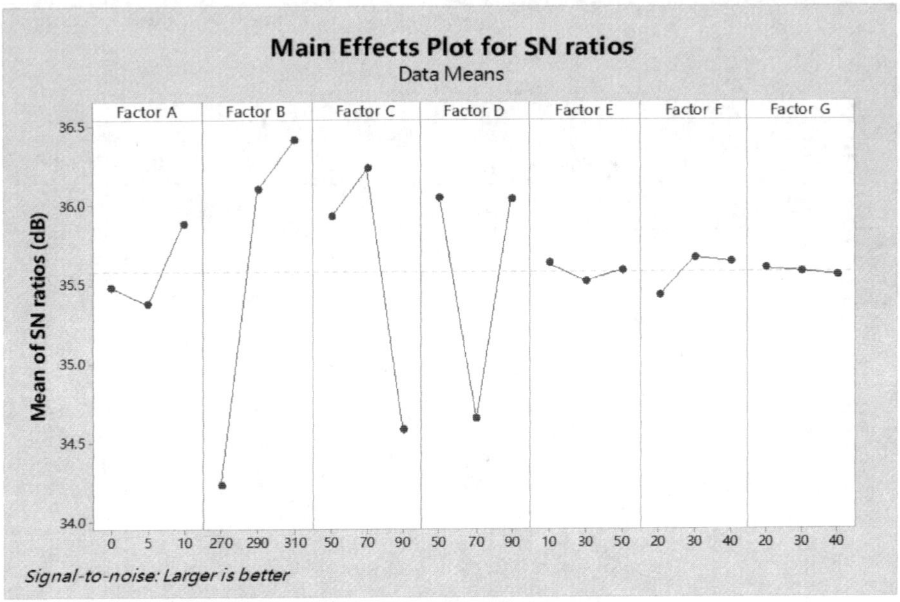

Figure 2. The effect of each factor on tensile strength of weld line.

IOP Conf. Series: Materials Science and Engineering **842** (2020) 012018 doi:10.1088/1757-899X/842/1/012018

Table 2. Mechanical properties of the tested specimens.

run	Tensile Strength (MPa)					Average	Standard deviation	*S/N* ratio
	Sample 1	Sample 2	Sample 3	Sample 4	Sample 5			
1	56.22	56.88	54.55	55.98	55.66	55.858	0.857	34.939
2	55.63	56.18	55.84	55.60	54.59	55.568	0.594	34.895
3	55.71	56.33	56.49	56.58	56.38	56.298	0.343	35.009
4	60.72	60.68	60.43	60.49	61.2150	60.706	0.307	35.664
5	60.86	61.68	61.35	61.86	61.40	61.430	0.381	35.767
6	60.22	61.02	61.79	61.02	61.61	61.132	0.616	35.724
7	60.05	61.29	60.86	60.15	59.11	60.292	0.836	35.603
8	61.65	62.40	62.23	61.48	62.15	61.982	0.396	35.845
9	61.87	61.00	62.67	62.91	61.98	62.086	0.751	35.858
10	62.81	61.59	60.18	62.06	61.96	61.720	0.968	35.806
11	59.73	56.89	55.15	55.19	55.86	56.564	1.905	35.039
12	56.26	51.65	53.87	53.29	52.88	53.590	1.700	34.571
13	56.60	58.92	57.48	57.96	56.66	57.524	0.967	35.194
14	59.20	59.97	59.02	59.83	59.88	59.580	0.437	35.501
15	58.70	59.28	58.28	59.83	58.38	58.894	0.653	35.400
16	60.54	58.21	60.08	59.60	58.57	59.400	0.988	35.473
17	62.07	59.01	60.84	60.53	61.08	60.706	1.110	35.661
18	61.05	60.48	62.67	61.39	60.57	61.232	0.885	35.737
19	48.54	42.69	43.21	41.28	41.87	43.518	2.904	32.731
20	41.84	42.71	40.40	42.01	41.33	41.658	0.859	32.389
21	43.41	42.91	42.41	43.57	43.58	43.176	0.508	32.703
22	65.81	73.73	73.56	72.03	71.95	71.416	3.242	37.053
23	71.90	71.45	69.56	69.47	70.93	70.662	1.102	36.981
24	71.89	76.25	77.10	76.23	78.33	75.960	2.431	37.601
25	80.62	81.95	81.89	81.33	81.91	81.540	0.574	38.227
26	70.65	77.83	77.84	76.56	76.71	75.918	3.006	37.589
27	72.05	77.03	78.02	79.33	78.10	76.906	2.835	37.704

Table 3. Summary of ANOVA results for the weld-line strength.

Factor	Sum of squares	DOF	Variance	F value	Confidence	Rank
A	6.5663	2	3.2832	1.48473	76.9%	4
B	124.9790	2	62.4895	28.25930	100.0%	1
C	69.5459	2	34.7729	15.72518	100.0%	2
D	58.4092	2	29.2046	13.20705	100.0%	3
E	0.2957	2	0.1479	0.06687	6.5%	6
F	1.5218	2	0.7609	0.34411	29.0%	5
G	0.0427	2	0.0213	0.00965	1.0%	7
Error	238.8194	108	2.2113	S= 1.4870		
Total	12727.5178	134				

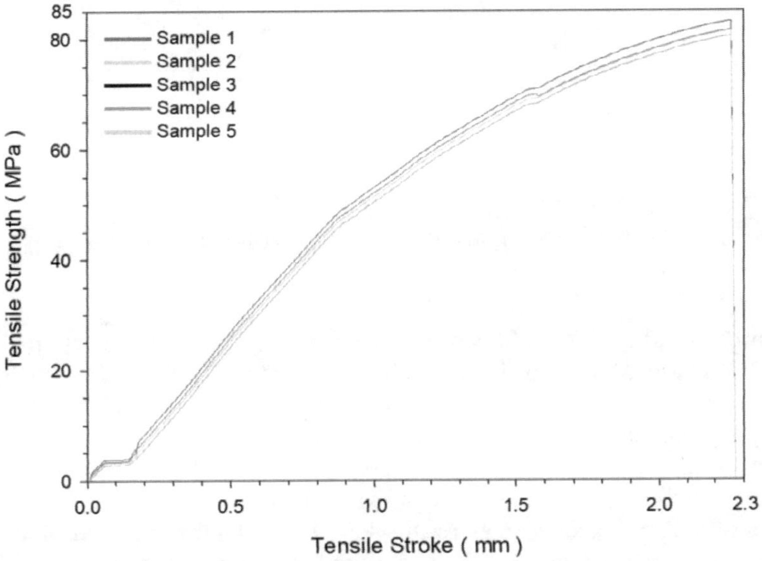

Figure 3. Tensile strength test of specimen made using Taguchi optimized parameters.

The Rank in Table 3 indicates that factor B (melt temperature) and factor C (mold temperature) are the important factors affecting the tensile strength of weld line. Therefore, other factors are fixed at the combination of A3-D1-E1-F2-G1, and the factors of B and C are adopted as the input variables. And then Minitab is used to plot the response surface as shown in Figure 4.

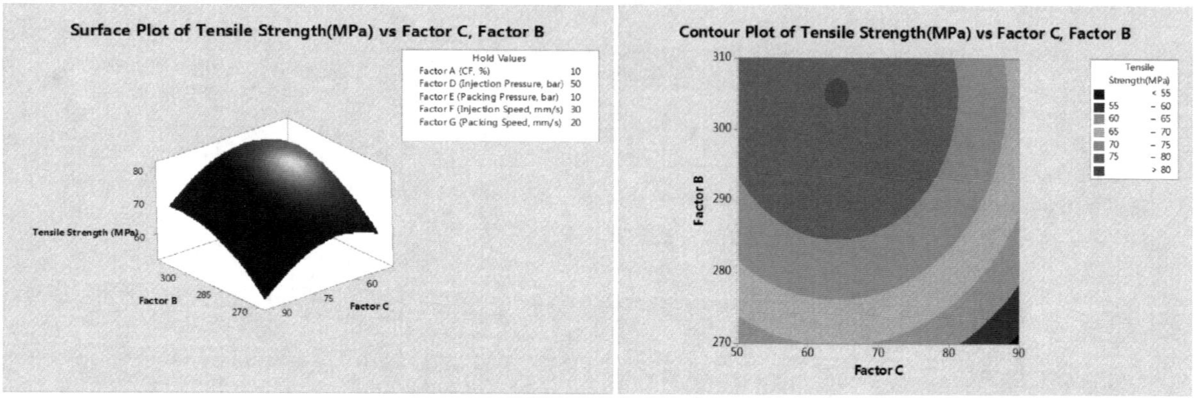

Figure 4. Response surface and contour plot of the tensile strength as a function of the melt temperature and the mold temperature.

The statistic software Minitab is used to establish the response surface method, and the tensile strength equation of non-interactive predicted under second-order model is:

$$\begin{aligned} \text{Tensile strength} &= \\ &-1020-0.777A+7.34B+2.178C-3.247D-0.117E+0.601F+0.055G+0.1282A^2 \\ &-0.01203B^2-0.01711C^2+0.02312D^2+0.00182E^2-0.00871F^2-0.00122G^2 \end{aligned} \quad (3)$$

The search of optimal solution of tensile strength equation will lead to the optimal combination of process parameters (Figure 5) of tensile strength, which includes CF 10%, melt temperature of 305.55°C, mold temperature of 64.54 °C, injection pressure of 50 bar, packing pressure of 10 bar, injection speed of 40 mm/s, and packing speed of 20 mm/s. The result of response surface method is approaching to the result of Taguchi method, while the results of these two analyses of factor F (injection speed) are different. However, the significance of factor F is lower, so it can be neglected. The comparison between the experimental results based on Taguchi method and response surface method is as shown in Table 4, where there is a certain difference between the analysis results of factor B (melt temperature) and factor C (mold temperature). So it is speculated that Taguchi method can only be used to search for the local optimum at the levels, while the response surface methodology can be used to find the global optimum outside the levels.

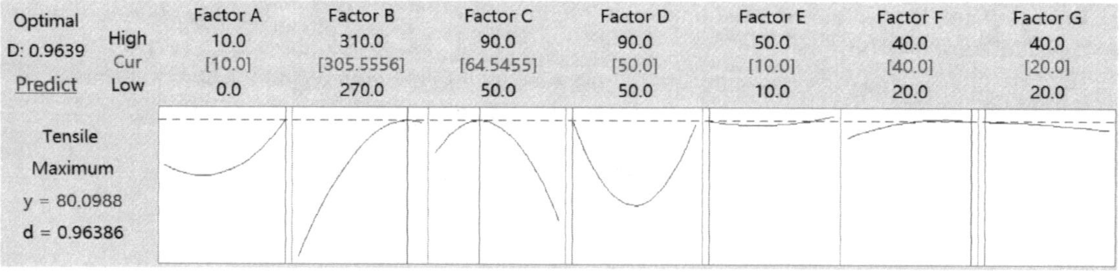

Figure 5. Optimized process parameters for maximum strength of weld line.

Table 4. Comparison of the optimized process parameters predicted using Taguchi method with that predicted by response surface method.

Factor	Taguchi method	Response surface method
A – Carbon fiber (%)	10	10
B – Melt temperature (℃)	310	305.55
C – Mold temperature (℃)	70	64.54
D – Injection pressure (bar)	50	50
E – Packing pressure (bar)	10	10
F – Injection speed (mm/s)	30	40
G – Packing speed (mm/s)	20	20

4. Conclusions
In this study Taguchi method is adopted to find the optimal process parameters in order to enhance the tensile strength of finished product, and then ANOVA is used to evaluate the errors in the experiment to enhance the credibility of experimental result. And then the response surface method is used for fine tuning of important process parameters, and it is verified that the optimal process parameters are obtained by Taguchi method. This study reveals that the factors affecting weld-line strength of polycarbonate are in the order of melt temperature, mold temperature, injection pressure, fiber content, injection speed, packing pressure, and packing speed. The tensile strength of weld line will be increased along with the increase of melt temperature. This is because that the increase of melt temperature will reduce the material viscosity, and there will be sufficient time for molecular entanglement to result in enhanced bonding rate. The result of this experiment can serve as the process improvement method to enhance the mechanical strength of plastic product, and it can be used to replace the metal to achieve the goal of weight reduction.

5. Acknowledgments
This work was supported by the Ministry of Science and Technology in Taiwan through the grants MOST 108-2628-E-008-004-MY2.

6. References
[1] E.M. Hagerman, 1973. Weld-line fracture in molded parts, *Plastics Engineering*, Vol. 29, No. 10, pp. 67-69.
[2] S.-G. Kim and N.P. Suh, 1986. Performance prediction of weld line structure in amorphous polymers, *Polymer Engineering and Science*, Vol. 26, No. 17, pp.1200-1207.
[3] C.-S. Chen, T.-J. Chen, R.-D. Chien, S.-C. Chen, 2007. Investigation on the weldline strength of thin-wall injection molded ABS parts, *International Communications in Heat and Mass Transfer*, Vol. 34, pp.448-455.
[4] B.N. Nguyen, L.S. Fifield, S.A. Kijewski, M.D. Sangid, J. Wang, X. Jin, F. Costa, C.L. Tucker III, R.N. Mathur, U.N. Gandhi, S. Mori, 2015. Predictive engineering tools for injection-molded long-carbon-fiber thermoplastic composites, FY 2015 First Quarterly Report, PNNL Report under Contract DE-AC05e76RL01830, The US Department of Energy, Pacific Northwest National Laboratory, PNNL-24031.
[5] X. Yan, S. Cao, 2018. Structure and interfacial shear strength of polypropylene-glass fiber/carbon fiber hybrid composites fabricated by direct fiber feeding injection molding, *Composite Structures*, Vol. 185, pp.362-372.
[6] D.C. Montgomery, 2009. *Design and analysis of experiments*, 7th ed., John Wiley & Sons (Asia), Hoboken, New Jersey.
[7] F.A. Radini, M. Ghozali, O. Ujianto, A. Haryono, 2019. Effect of fillers and processing parameters on the shrinkage of injected molding polyamide 66, *International Journal of Materials, Mechanics and Manufacturing*, Vol. 7, pp.165-169.

IOP Conf. Series: Materials Science and Engineering **842** (2020) 012019 doi:10.1088/1757-899X/842/1/012019

Research on Improving the Taper of Hole Processing by Insulation Coating of ECM Cathode

Jianxiao Bian[1,2,*], Baoji Ma[2], Lijun Qi[2], Honghong Gao[2]

[1] School of Mechanical Engineering ,Longdong University , Qingyang 745000, China
[2] School of Mechanical Engineering ,Xi'an Technological University , Xi'an 710021, China
*E-mail:Jxbian@ldxy.edu.cn

Abstract. Electrochemical machining (ECM) is widely used in metal parts processing because of its high processing speed, good surface quality, no stress and deformation of the workpiece after processing, and no tool loss. However, when using electrochemical drilling, the taper of the hole is large due to the secondary electrolysis phenomenon. Therefore, this paper aims to reduce the taper of the anode workpiece hole by studying the degree of insulation of the cathode sidewall. The processing of the holes is performed by insulating coating the cathode structures to form different conductive radial heights h_0. A multiphysics (electric field, flow field, temperature field and mathematical field) model for simulating ECM machining process using a moving cathode tool is proposed in different modes of conductive radial height h_0. The mathematical relationship between the height h_0 and the taper of the machined hole is established to obtain the optimum machining method.

1. Introduction

Electrochemical machining (ECM) is widely used in metal parts processing because of its high processing speed, good surface quality, no stress and deformation of the workpiece after processing, and no tool wear[1]. However, ECM has the problems of high taper and low efficiency in the processing of pores due to secondary electrolysis[2-3]. Therefore, how to use ECM high-quality and efficient processing holes has become a research hotspot in many countries.

Since electrolytic machining is to re-engraving the surface structure of the cathode material, the design of the cathode has a great influence on the improvement of the processing quality of the hole[4]. The use of sidewall insulating electrodes in the ECM process can improve the processing accuracy of the holes[5]. The sidewall insulation can be divided into full insulation and partial insulation, and the relationship between the degree of partial insulation and the taper of the formed profile has not been studied. In order to reduce the taper of the hole, Rolf Schuster[6] proposed using ultrashort voltage pulses to locate the processing area during electrochemical micromachining and to process some 3D microstructures on the copper plate. Fang X.L et al.[7] applied a potential difference between the auxiliary electrode and the anode to improve the accuracy of the micropore. Kim et al.[8] used disc electrodes to machine microporous structures and used array cathodes to improve processing efficiency. Yong L et al.[9] studied the ultra-short voltage pulse and high-speed rotating spiral electrode in the electrochemical drilling process to produce non-tapered small holes, and obtained the influence of process parameters on machining accuracy. In addition, in order to improve the processing quality of the holes, various electrochemical composite processing methods have been studied.

This paper aims to reduce the taper of the anode workpiece hole by studying the degree of insulation of the cathode sidewall. The processing of the holes is performed by insulating coating the cathode structures to form different conductive radial heights h_0, as shown in Fig. 1. Let ΔS_x approach X_0 to reduce the taper of the hole. A multiphysics (electric field, flow field, temperature field and mathematical field) model for simulating ECM machining process using a moving cathode tool is proposed in different modes of conductive radial height h_0. And establish the mathematical relationship between h_0 and the taper of the machined hole to get the best processing method.

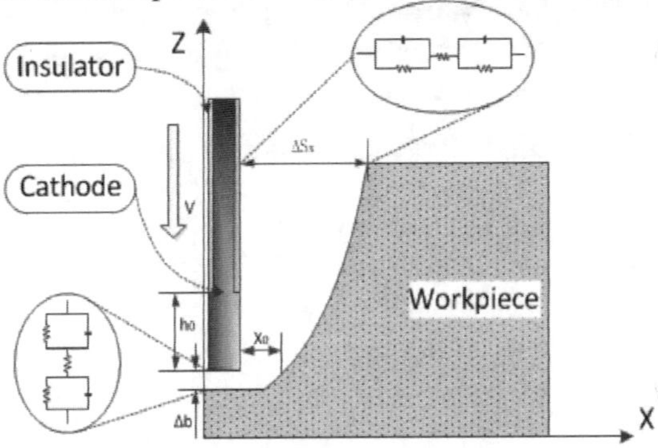

Figure 1. ECM processing of pores.

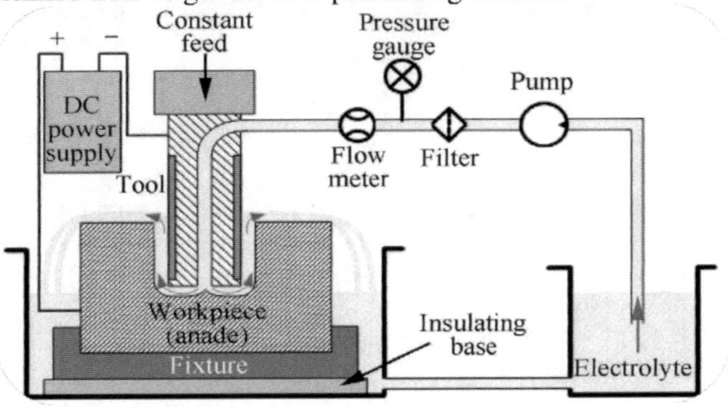

Figure 2. Electrochemical processing schematic.

2. Models

ECM is a processing technology that uses the principle of metal anode dissolution to complete the etching of the workpiece. During the processing, the negative pole of the power supply is connected with the tool, the positive pole of the power supply is connected with the workpiece to turn on the pulse current, and the electrolyte flows at high speed between the two poles, at the same time keep the cathode tool at a certain speed. According to Faraday's law, the anode workpiece is continuously dissolved according to the shape of the tool of the cathode, so that the size and shape of the workpiece are re-engraved to achieve the processing purpose. Among them, the cathode is transformed from a solid to a hollow structure due to the design of the flow channel. as shown in Fig. 2.

2.1. Geometry model

In order to study the effect of introducing an insulated coated cathode on the machining accuracy, a physical model was constructed, see in Fig. 3. Among them, the cathode boundaries are 3, 4, 5, 7, 8, and 9. Where boundaries 3 and 9 can be partially electrically insulated from the electrolyte. The anode boundary is 12 and the entity is an electrolyte that flows between the two poles. The cathode has a diameter of 10 mm and the hollow structure has a diameter of 2 mm. The interelectrode gap (IEG) is 0.2 mm. The cathode moves downward at a constant rate V during processing. A constant potential is applied to the surface of the anode in contact with the electrolyte, and the cathode is grounded. The electrolyte flows in a radial direction at a uniform rate, i.e., flows in from AB, and flows out from CD and EF.

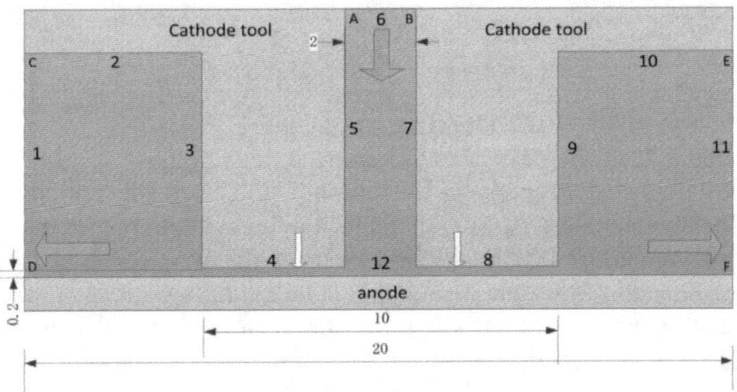

Figure 3. Geometry model.

IOP Conf. Series: Materials Science and Engineering **842** (2020) 012019 doi:10.1088/1757-899X/842/1/012019

2.2. Theoretical model

According to the electric field theory, the potential distribution is given by the Laplace equation:

$$\nabla^2 \varphi = 0 \tag{1}$$

Where, φ is the potential.

The relationship between current density J and potential φ is derived from Ohm's law.

$$J = k \frac{\partial \varphi}{\varphi n} \tag{2}$$

Where, n is the unit vector perpendicular to the workpiece surface; k is the conductivity of the electrolyte.

Because the motion of the flow field is considered as "turbulence", the velocity distribution of electrolyte in the simulation is represented by the standard model, which is called N-S equation by using continuity equation and momentum equation.

$$\frac{\partial \rho}{\partial t} + \frac{\partial}{\partial x_j}(\rho u_j) = 0 \tag{3}$$

$$\frac{\partial}{\partial t}(\rho u_t) + \frac{\partial}{\partial x_j}(\rho u_j u_i) = \frac{\partial}{\partial x_j}(-p\delta_{ij} + \mu_{eff}(\frac{\partial u_i}{\partial x_j} + \frac{\partial u_j}{\partial x_i})) + F_i \tag{4}$$

where , $i, j = 1,2,3$; ρ is the fluid density ; t is the time ; u_i, u_j are the fluid velocity component; x_j is the coordinate component; F_i is the volume force component; σ_{ij} is the stress tensor; p the pressure; μ the molecular viscosity; δ_{ij} is the unit tensor, when $i \neq j$, $\delta_{ij} = 0$, when $i = j$, $\delta_{ij} = 1$.

Where, μ_{eff} represents the effective viscosity and we have $\mu_{eff} = \mu + \mu_T$, μ_T is turbulent viscosity

$$k: \quad \frac{\partial}{\partial t}(\rho k) + \frac{\partial}{\partial x_j}(\rho u_j k) = \frac{\partial}{\partial x_j}\left[(\mu + \frac{\mu_T}{\sigma_k})\frac{\partial k}{\partial x_j}\right] + G - \rho\varepsilon \tag{5}$$

$$\varepsilon: \quad \frac{\partial}{\partial t}(\rho k) + \frac{\partial}{\partial x_j}(\rho u_j k) = \frac{\partial}{\partial x_j}\left[(\mu + \frac{\mu_T}{\sigma_\varepsilon})\frac{\partial \varepsilon}{\partial x_j}\right] + \frac{\varepsilon}{k}(C_{\varepsilon 1}G - C_{\varepsilon 2}\rho\varepsilon) \tag{6}$$

According to Faraday's law, the removal rate of the anode workpiece is

$$v = \eta \omega J \tag{7}$$

Where, η is the current efficiency; ω is the chemical equivalent; J is the current density.

3. Simulation

3.1. Simulation Parameters

It can be seen from formula 7 that the value of radial height h_0 can affect the current density distribution in the inner wall of the hole and further affect the anodic dissolution. Change the outline of its processing. The purpose of this paper is to study how to reduce the taper in hole electrochemical machining by improving the traditional cathode. Therefore, in order to obtain the relationship between the radial height h_0 and the current density and the shape of the anode profile, we studied the changes of current density and temperature in the electrochemical machining process with different values of h_0. And the relationship between them and the processing profile is established. The physical field of coupling simulation is electric field, flow field and temperature field. The parameters and values of the simulation model are shown in Table 1. The conductivity of electrolyte is measured by DDS-307 Conductivity Meter (Instrument and Electrical Science Instruments Co., Ltd., Shanghai, China). fig. 4 is a flow chart for simulation of anode profile forming.

Table 1. Simulation model parameters and value.

Model parameters	values
Potential difference (V)	10
IEG value (mm)	0.2
Electrolyte electric conductivity(S/m)	1.542
Thermal conductivity(W/m*k)	0.58
Temperature reference(℃)	20
Inlet flow rate of electrolyte(m/s)	0.5
Electrolyte composition	1mol/L$NaNO_3$+0.1mol/L$C_6H_8O_7$
Radial height(mm)	0,0.6,1.2,1.8,2.4, 3

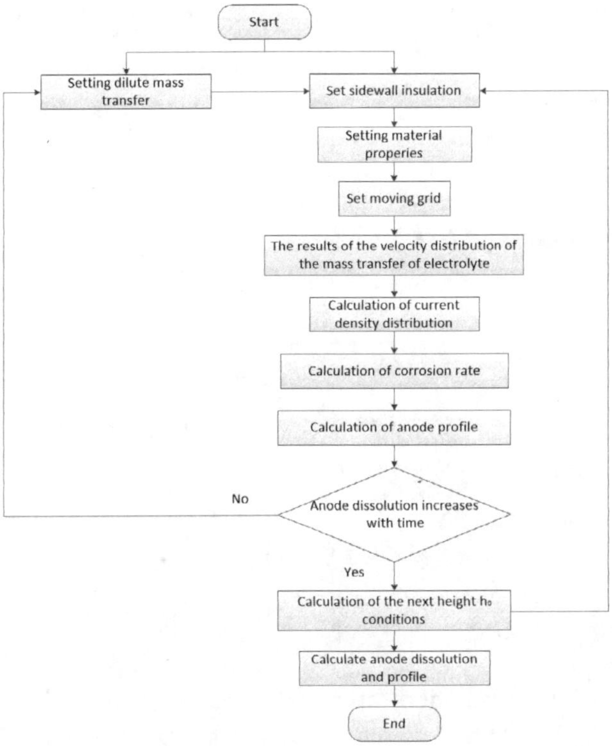

Figure 4. Flow chart for simulation of anode profile forming.

3.2. Simulation results

3.2.1. Temperature distribution. In ECM, the temperature distribution is mainly affected by electrolyte convection, and the velocity at the entrance is set to be constant and vertical downward. Fig. 5 shows the distribution of temperature and velocity after 60 seconds of anodic dissolution when h_0 is 3 mm. It can be seen that the higher temperature is mainly concentrated in the gap of processing. Fig. 6 shows the temperature distribution after 60 seconds of anodic dissolution at different h_0. It can be seen that the minimum temperature is 303K when h_0 is 0.6 and 1.8mm. For other h_0 values, the gap temperature changes slightly. As a whole, the temperature changes in a small range with the change of the insulation height of the side wall. It has little effect on the accuracy of processing.

IOP Conf. Series: Materials Science and Engineering **842** (2020) 012019 doi:10.1088/1757-899X/842/1/012019

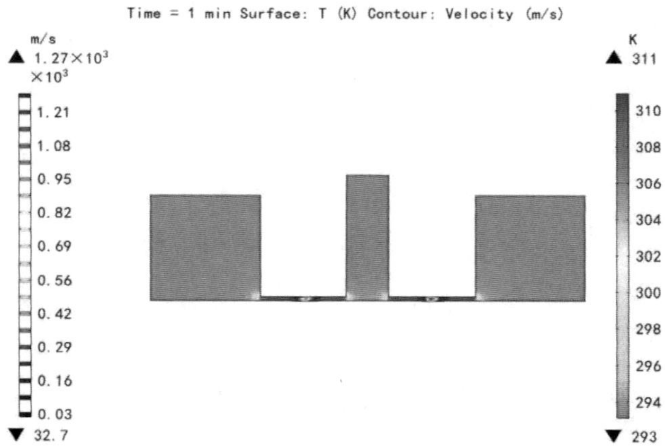

Figure 5. Temperature distribution of electrolyte after 60 seconds of anodic dissolution when h_0 is 3 mm.

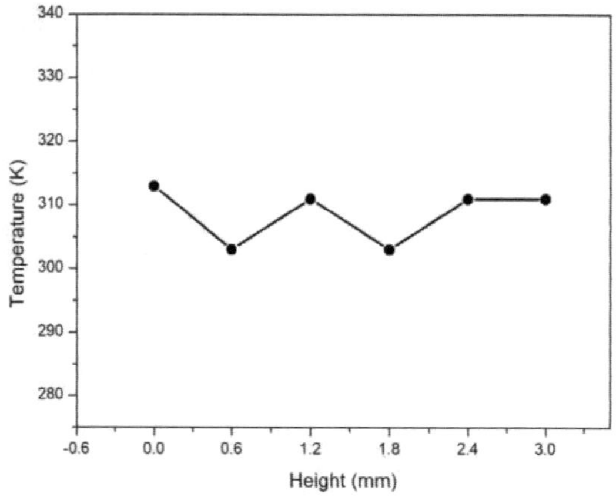

Figure 6. Temperature distribution of electrolyte under Different h_0 Conditions.

3.2.2. Current density distribution. Current density has the greatest impact on the accuracy of electrochemical machining, which directly affects the dissolution of the anode. Fig. 7 shows the current density and potential distribution of electrolyte for 60s of anodic dissolution when h_0 is 3 mm. Fig. 8 shows the current density distribution of electrolyte for 60s of anodic dissolution under different h_0. It can be seen from the fig. 8 that the current density of electrolyte for 60s of anodic dissolution is the highest when h_0 is 1.2 mm.

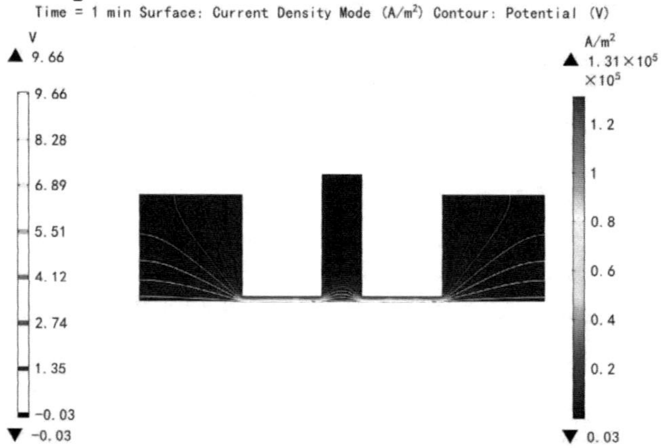

Figure 7. Current density distribution after 60 seconds of anodic dissolution when h_0 is 3 mm.

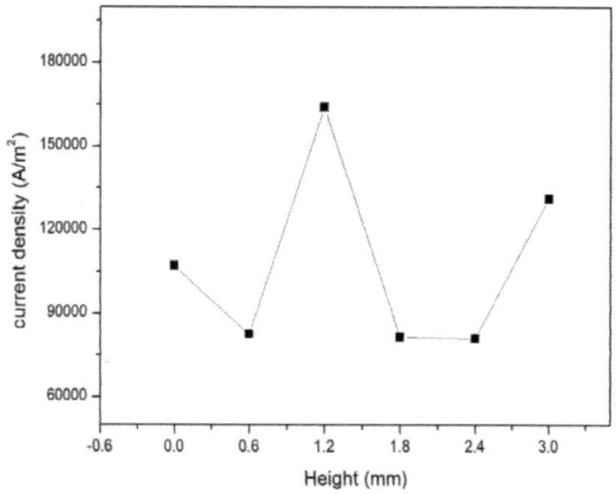

Figure 8. Current density distribution of electrolyte under Different h_0 Conditions.

3.2.3. Anode profile forming distribution. As shown in Fig. 9, the anode forming profile simulated by the ALE moving grid at different h_0, it can be seen from the figure that as the h_0 continues to increase, the taper of the anode hole is continuously reduced. In order to obtain the mathematical model of the hole machining taper and h_0, the data is fitted through the matlab polynomial function model and the function relationship is obtained. The resulting mathematical model can quantitatively determine the relationship between h_0 and the taper of the machined hole.

$$Y = 0.13589 \times h_0^{3} - 2.486 \times h_0^{2} + 13.13 \times h_0 + 9.0059 \tag{8}$$

Wherein, Y is the taper of the hole.

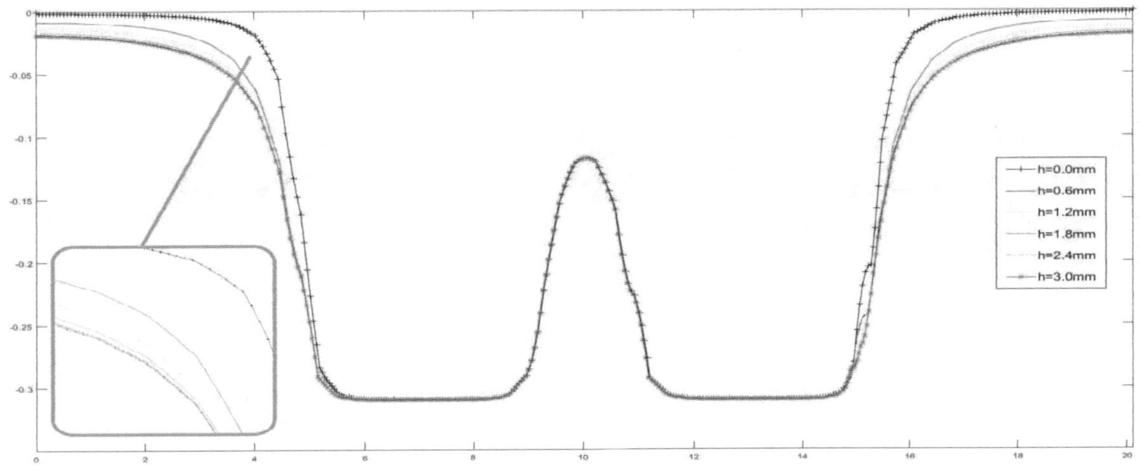

Figure 9. Anode forming profile simulated with ALE moving grid at different h_0.

4. Conclusions

This paper presents a cathode design method to improve the taper of electrolytic machining holes. The simulation of multi-field coupling is carried out, and the mathematical relationship between hole taper and cathode sidewall insulation is quantitatively described. The conclusions are as follows:

- Propose a partial insulation of the sidewall of the ECM cathode to machine the hole to reduce its taper.
- Establish a multiphysics model that uses mobile cathode tools to simulate ECM processing.
- Obtain a mathematical model between the different insulation heights of the cathode sidewalls and the taper of the processing holes, which can quantitatively reflect the relationship.

Acknowledgments

This study was supported by Doctoral Foundation Project of Longdong University(Grant No. XYBY1905).

Reference

[1] Li L, Baoji MA. Effect of magnetic field on anodic dissolution in electrochemical machining[J]. Int J Adv Manuf Technol, 2018, 94(1-4): 1-11.

[2] Li L, Baoji MA, Wang R, Du L. The coupled effect of magnetic field, electric field, and electrolyte motion on the material removal amount in ECM[J]. Int J Adv Manuf Technol, 2017, 91: 2995-3006.

[3] Xu Z, Sun L, Hu Y, Zhang J. Flow field design and experimental investigation of electrochemical machining on blisk cascade passage[J]. Int J Adv Manuf Technol, 2014, 71(1-4):459-469.

[4] Fang X, Zou X, Zhang P. Improving machining accuracy in wire electrochemical micromachining using a rotary helical electrode[J]. Int J Adv Manuf Technol Technology, 2016, 84(5-8): 929-939.

[5] Liu G, Li Y, Chen X, Ma X. Zhu, X. Side-Insulation of Tool Electrode and Application in Micro ECM[J].Electromach. Mould, 2009, 4:28-31.

[6] Schuster R, Kirchner V, Allongue P, Ertl G. Electrochemical micromachining[J]. Science, 2000, 289:98-101.

[7] Fang XL, Qu NS, Zhang Y, Zhu D. Improvement of hole exit accuracy in electrochemical drilling by applying a potential difference between an auxiliary electrode and the anode[J]. J. Mater. Mach. Technol, 2014, 214: 556-564.

[8] Kim BH, Na CW, Lee YS, Choi DK, Chu CN. Micro electrochemical machining of 3D micro structure using dilute sulfuric acid[C]. CIRP Ann. Manuf. Technol, 2005, 54:191-194.

[9] Liu Y, Mh L, Jr N, Sz L, Jiang Y. Fabrication of Taper Free Micro-Holes Utilizing a Combined Rotating Helical Electrode and Short Voltage Pulse by ECM[J]. Micromachines, 2019, 10(1):28

IOP Conf. Series: Materials Science and Engineering **842** (2020) 012020 doi:10.1088/1757-899X/842/1/012020

Early Warning of Position Accuracy for Pin Hole using Rules and Model based on LSSVM and PSO during Tractor Manufacturing Process

Chao Jiang, Ying Li, Zhilin Zhao, Fang Cheng*

College of Biosystems Engineering and Food Science, Zhejiang University, 310058, China

*E-mail: fcheng@ zju.edu.cn

Abstract. The factors affecting the processing quality of pin hole on the end face of tractor gearbox were analyzed. The main influencing factors on position accuracy of hole group include ambient temperature, the number of workpieces processed by drilling cutter and so on during precision manufacturing. The early warning rules of processing quality during tractor manufacturing process were proposed based on single-factor analysis and double-factors analysis. Then an early warning model for position accuracy of pin hole was established based on LSSVM while PSO was used to optimize the unknown parameters. The experimental results show that the RMSE of the model is 0.0058 and the RE is 0.075%, which verifies the validity of the established processing quality early warning model.

1. Introduction

Tractor is typical agricultural machinery and the power of agricultural production. The improvement of tractor's manufacturing quality will promote the development of agriculture. The driveline is a key component of tractor. The front and rear gearbox is connected by positioning the pin holes on the joint surface to form the transmission system. The position accuracy of pin holes on the end face of gearboxes contributes a lot to the quality of the connection, which further affects the overall performance of the tractor. Effectively controlling the position accuracy of pin hole is the key to ensure the manufacturing quality of the drive train of tractor.

Enterprises of tractor manufacturing usually adopt strategy of first article inspection or sampling inspection to control the quality of workpieces, in which way they can't obtain the quality data of every piece and have to use a post-error prevention mechanism. As to the position accuracy of pin hole on the end face of tractor gearbox, only the first piece in a batch is tested by the three-coordinate measuring instrument located in laboratory. The measurement of the hole diameter is usually made by plug gauges or callipers on site in manufacturing enterprises. The online measurement technology and early warning method of position accuracy remain to be explored to meet the requirements of quality control.

Precision manufacturing is the key to improve the level of the manufacturing industry. For precision manufacturing of the pin hole on end face of tractor gearbox, the position accuracy is highly demanded whereas affected by many factors. Method of early warning is needed especially to avoid damage because it is difficult to acquire intermediate quality data in a closed manufacturing environment using CNC （computer numerical control） machine tool. Research on the early warning rules and models for precision manufacturing of the hole group will certainly help to improve manufacturing quality and reduce manufacturing costs.

Early warning means giving notices or warnings and taking measures before a disaster or danger occurs, according to previous experience, knowledge or model, to prevent hazards and reduce unnecessary losses. The quality early warning modelling refers to the construction of quality control model according to relevant theory and the law of the occurrence of the hazard. The passive quality control is thus changed to the active quality control [1].

Steps used for constructing quality early warning model include analysis of influencing factors on processing quality and then algorithms such as support vector machine (SVM) [2], neural network, grey prediction model [3] chosen for modelling. Some commonly used methods for optimizing the early warning model are particle swarm optimization algorithm (PSO) [4], differential optimization algorithm [5], and colony algorithm [6], etc.

Subtractive clustering fuzzy recognition and fuzzy reasoning algorithm were used to construct the quality prediction model of deep hole surface roughness considering the speed of cutting and feeding as the influencing factors [7]. Machining parameters such as tool radial forward angle, tool nose radius, cutting speed and feed rate were drawn as the influencing factors of surface machining quality and analyzed to construct model using genetic algorithm [8].

The prediction model of turning quality during the production of automobiles was established based on neural network and particle swarm optimization considering feed speed, tooltip radius, cutting speed and depth as CNC turning process parameters [9]. An early warning model was established based on SVM whose unknown parameters were determined combining firefly algorithm and random heuristic search algorithm to optimize [10].

Based on analyzing the factors affecting the processing quality of pin hole on the end face of tractor gearbox, the objective of the research is to establish the early warning rules and intelligent model of position accuracy during precision manufacturing process of hole group.

2. Knowledge acquisition and early warning rules

2.1. Influencing factors on processing quality

The manufacturing quality of parts affects the final quality of products directly. During the manufacturing process, the quality of parts is affected by many factors. According to the source of the factors, it can be divided into human, machine, material, method, environment and measurement.

The main influencing factors of gearbox processing quality are machine tool fixture and spindle rotation error caused by ambient temperature, tool deformation caused by long working time, tool wear and fixture deformation caused by machine tool working time, etc. Especially for high precision manufacturing, the influence of ambient temperature on the quality of processing cannot be ignored. As to the working time of tool, the number of workpieces processed by drilling cutter is considered as an influencing factor because reamer will change the cutter after processing certain pieces. The diameter of pin hole interacts with the position accuracy because they are formed under the same conditions.

2.2. Analysis of influencing factors

Practically data of pin hole on the end face of gearboxes were collected from manufacturing execution system (MES) using in tractor enterprise to investigate the correlation between position accuracy with its main influencing factors. Since the influencing factors and the method of establishing quality early warning model of horizontal position accuracy are the same as vertical position accuracy, vertical position accuracy was taken as a representative example to analyze the influencing factors and establish quality early warning rules and model in the following sections. The available data of position accuracy is limited because the strategy of first article inspection mentioned above. Typical vertical position accuracy measured by three-coordinate measuring instrument during 10 months in 2018 was analyzed using SPSS software correlation analysis method. The results of correlation analysis of single factors are shown in Table 1. It verified that the three factors of ambient temperature, number of workpieces processed by drilling cutter and diameter of pin hole are correlated with position accuracy respectively.

Table 1. Relevance Analysis.

Relevance	x_1-Ambient temperature (℃)	x_2-Number of workpieces processed by drilling cutter	x_3- Diameter of pin hole (mm)
y-Position accuracy (mm)	0.673	0.435	0.712

MATLAB software was used to do double-factors analysis. The data fitting maps are shown in Fig.1.

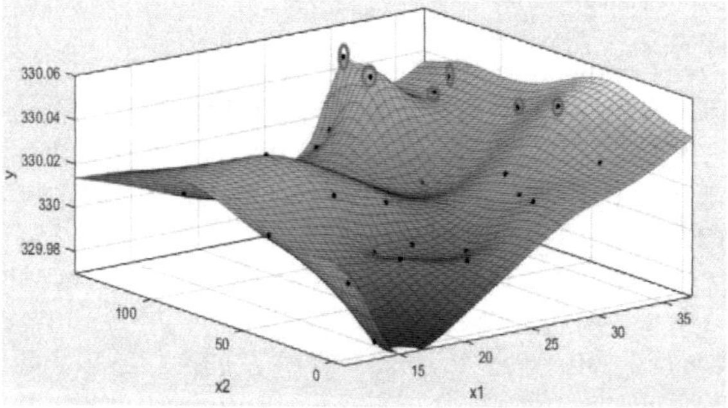

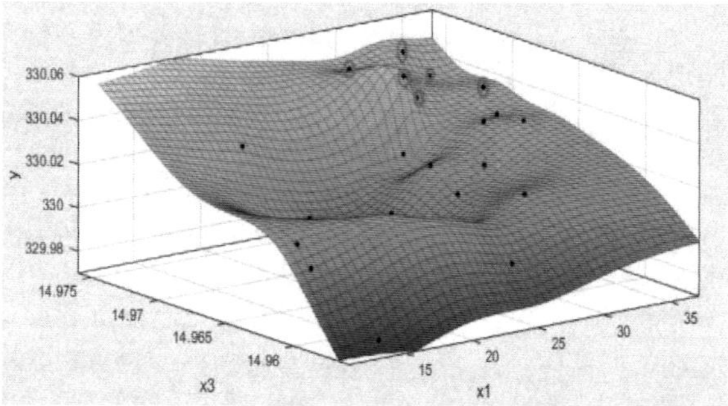

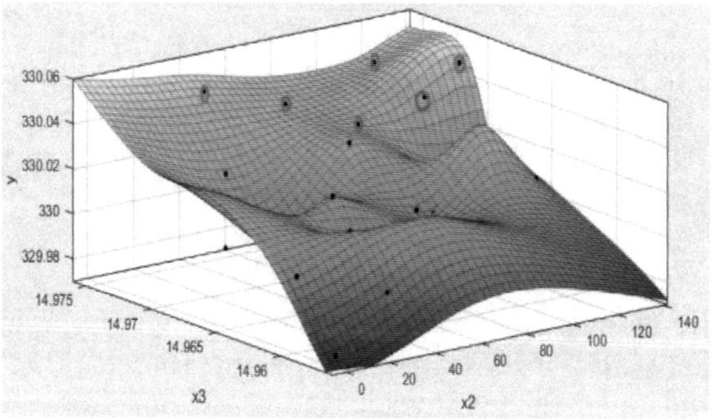

x_1: Ambient temperature (℃)
x_2: Number of workpieces processed by drilling cutter
x_3: Diameter of pin hole (mm)
y : Position accuracy (mm)

Figure 1. Data fitting diagram where red marks represent the over-tolerance data of position accuracy.

2.3. Early warning rules

The early warning rules are proposed as follows based on above single-factor analysis and double-factors analysis.

1. The ambient temperature exceeds 30 degrees Celsius and more than 120 pieces were processed by drilling cutter.
2. The ambient temperature exceeds 30 degrees Celsius and the diameter of pin hole is greater than 14.970 mm.
3. The diameter of pin hole is greater than 14.970 mm and more than 120 pieces were processed by drilling cutter.

When either one of the above three rules is met, the position accuracy is more likely to be over-tolerance, and error prevention mechanism should be taken, such as checking machine tool fixture and spindle rotation, changing machine tool cooling fluid, or replacing drill cutter, to avoid potential over-tolerance of position accuracy for following gearbox manufacturing.

3. Early warning model based on LSSVM and PSO

3.1. Least squares support vector machine (LSSVM)

SVM is a statistical learning theory algorithm. It can also find the global optimal solution for small sample size, non-linearity and high dimension pattern recognition problems with high prediction accuracy. Because of the introduction of support vector, it reduces the computational complexity and avoids over-learning of models. SVM can be used in classification, regression prediction and early warning of engineering applications. It is widely used, but it is a quadratic programming problem, which is prone to multiple local optima. To solve this problem, Suyken et al. proposed LSSVM, which transforms quadratic programming into linear programming. Compared with SVM, LSSVM has a simpler structure and fast calculation speed.

The basic idea of SVM is to transform data from low-dimensional space to high-dimensional space using kernel function and to minimize the sum of squares of distances from data to the hyperplane. LSSVM use least square method to transform inequality constraints into equality constraints.

$$J(w,\xi)=\min_{w,b}\frac{1}{2}\|w\|^2+\frac{1}{2}C\sum_{i=1}^{N}\xi_i^2$$
$$s.t.\quad y_i=w^T\varphi(x_i)+b+\xi_i\quad i=1,2,\cdots N$$

Where C is a penalty factor, ξ_i is a relaxation variable and $J(w,\xi)$ is an optimization function. The quality early warning model base on LSSVM is deduced and calculated as follow.

$$f(x)=\sum_{i=1}^{N}\alpha_i K(x,x_i)+b$$

Radial basis function shown below was used as the kernel function of LSSVM.

$$K(x_i,x_j)=\exp\left[\frac{-(x_i-x_j)^2}{2\sigma^2}\right]$$

Therefore, the key to establish the quality early warning model is to find the parameters σ of the radial basis function and penalty factor *C*.

3.2. Particle swarm optimization (PSO)

PSO is an optimization algorithm proposed to simulate the feeding process of birds. It can be used to find the optimal solution of function. PSO can be used to optimize function, train neural network and so on under no need to set too many parameters, while achieve fast convergence speed and good ability to find the global optimal solution.

PSO uses massless particles to simulate birds' foraging. Particles have two characteristics: velocity and location, which represent the moving speed and location of each particle. All particles search for the optimal solution in their search space, share their optimal solution with all particles and compare to know the current global optimal solution. Then the particle changes its speed and location through the current optimal solution. In this research, PSO was used to find the global optimal solution of parameters σ of radial basis function and penalty factor C.

4. Results

Set the number of particle swarm *N*=20, velocity of particle swarm $v_{min}=v_{max}=25$, learning factors *C1*=*C2*=2, weight between 0.2 and 0.5, maximum number of iterations =200. The results of global optimal solution of parameters of radial basis function and penalty factor are σ=0.01 and *C*=113.7015 respectively.

Quality early warning model of pin hole position accuracy was established based on LSSVM and PSO. Then the early warning model was verified using vertical position accuracy data of pin hole on the end face of tractor gearbox, which were measured with a three-coordinate measuring instrument in a tractor processing workshop from February to November 2018.

Considering the different measurement units of the collected data include temperature, number of workpieces and diameter, the normalization method was used to pre-process the data at first. To obtain as much information as possible from experimental data and avoid over-fitting, k-fold cross-validation method was used to test and train the collected data. The realization process of the k-fold cross-validation method is briefly introduced as follows.

Step 1: Divide the normalized data into k parts randomly.

Step 2: Select one of them as test set and the rest as training set.

Step 3: Make every data set has an opportunity to be a test set, calculate the model from each training set, and test with the test set. This paper uses the average absolute error between the test value and the actual value as the evaluation index.

Step 4: At the end of training, the average value of k measurement parameters is calculated and used as the precision index of the model.
Then root-mean-square error (RMSE) and relative error (RE) were calculated to analyze and evaluate the effectiveness and stability of the quality early warning model. Results show that RMSE of the quality early warning model is 0.0058 mm and RE is 0.075%.

5. Conclusion

The research focuses on the issues that high-precision processing exits the problem of the position accuracy of pin hole affected by ambient temperature and other factors, while it is difficult to obtain quality inspection data in real time during manufacturing processes using numerical control machine. Firstly, the early warning rules of pin hole position accuracy were putted forward in this paper according to acquired knowledge and factors analysis, which can be used for qualitative warning. Then an early warning model based on LSSVM and PSO was constructed to estimate the processing quality of pin hole position accuracy quantitatively in advance. The experimental results show that the RMSE and RE of the quality early warning model are 0.0058mm and 0.075% respectively. So it can be concluded that the quality early warning model is effective and can provide decision support information for subsequent optimization of processing quality, improve product quality and reduce production cost.

Acknowledgments

The research was funded by the National Key R&D Program of China (2017YFD0700205). Author thanks for the support of CSC scholarship.

References

[1] Zhai CL. (2017). Risk early-warning model of power transmission network based on landslide disaster [D]. *Chongqing University*.

[2] Amari S., Wu S. (1999). Improving support vector machine classifiers by modifying kernel functions. *Neural Networks,* 12 (6): 783-789.

[3] Lin C., Chen Y. (2018). Grey optimization of injection molding processing of plastic optical lens based on joint consideration of aberration and birefringence effects. *Microsystem Technologies*, 25 (2): 621-631.

[4] Wen-Fung L., Yen G.-G. (2008). PSO-based multiobjective optimization with dynamic population size and adaptive local archives. *IEEE Transactions on Systems Man & Cybernetics Part B,* 38 (5): 1270-1293.

[5] Storn R., Price K. (1997). Differential Evolution – A Simple and Efficient Heuristic for global Optimization over Continuous Spaces. *Journal of Global Optimization*, 11 (4): 341-359.

[6] Zhang JH., Gao QS. (2000). A Self-Adaptive Ant Colony Algorithm. Control Theory & Applications, 17 (1): 1-3.

[7] Alata M., Demirli K. (2004). Prediction model for BTA deep-hole machining using fuzzy clustering approach: experimental study. *Materials and Manufacturing Processes*, 19 (6): 1103-1119.

[8] Suresh Kumar Reddy N., Venkateswara Rao P. (2005). Selection of optimum tool geometry and cutting conditions using a surface roughness prediction model for end milling. The International *Journal of Advanced Manufacturing Technology*, 26 (11-12): 1202-1210.

[9] Ahilan C., Kumanan S., Sivakumaran N., et al. (2013). Modeling and prediction of machining quality in CNC turning process using intelligent hybrid decision making tools. *Applied Soft Computing*, 13 (3): 1543-1551

[10] Tuba E., Mrkela L., Tuba M. (2016). Support Vector Machine Parameter Tuning using Firefly Algorithm[C]. *Radioelektronika*. 413-417.

IOP Conf. Series: Materials Science and Engineering **842** (2020) 012021 doi:10.1088/1757-899X/842/1/012021

Influence of Bearing's Flexibility on the Working of Cycloid Drive

Y V Sinitsyna[1] and M M Ermolaev[1]

[1]Fundamentals of Machine Design Department, Bauman Moscow State Technical University, 2-ya Baumanskaya st. 5, Moscow 105005, Russia

[a]J.Sinitcyna@gmail.com, [b]mmermolaev@gmail.com

Abstract. In this article are reviewed calculate method of force distribution with accounting bearing flexibility on cycloid reducer elements of the types KHV and 2KV. There are analyzed schemas such as gear without machining errors, with constant errors on the pins and with random errors from machining tolerance. Schemas of force distribution on the pins of the types KHV and 2KV are presented. Relation between bearings flexibility and force distribution on the pins are determined, and also relation between machining errors and force distribution on the pins. Advantages and disadvantages of various methods of load estimation in the reducer are elucidated, appliance of this methods for calculation of various types cycloid reducers are analyzed.

1. Introduction

Modern engineering requires that gears possess such properties as compactness, rigidity and accuracy. These characteristics correspond to the cycloid speed reducers, which is widely used as rotary axes of machines, manipulators [1,2] and other technological equipment. Achievement of such requirements is obtained due to the multiple-tooth contact, however, this leads to the fact that the cycloid reducer forms a statically uncertain system and, as a result, special requirements to the processing accuracy are applied and the force calculation becomes more complicated. The bearing rigidity makes a significant contribution to the accuracy of the reducer. Thus, the development of cycloid reducer new models, their optimization, production and implementation are quite expensive processes.

Let's consider the most promising cycloid reducers such as KHV and 2KV according to Kudravtsev [3]. The KHV gear (fig. 1.*a*) consist of input shaft with an eccentric *h*, which confers the cycloidal gear rotation movement *g*[4]. This movement is limited by the mechanism of parallel cranks w relative to the output shaft *v*.

IOP Conf. Series: Materials Science and Engineering **842** (2020) 012021 doi:10.1088/1757-899X/842/1/012021

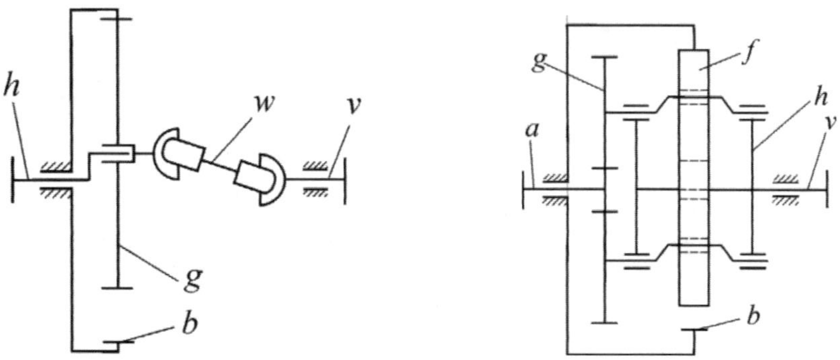

a. KHV gear b. 2KV gear

Figure 1. Kinematic schemas

The gear 2KV (fig. 1.*b*) consists of a center gear with involute tooths *a*, it engages with several spur gears *g*, which are positioned on eccentric shafts *h*, and cycloidal gear *f*, which engages with the pin wheel *b*, and carrier *v*. Given that are several eccentric shafts, they work like parallel cranks.

2. Calculation methods

The existing method of calculation assumes that all the details of the cycloidal reducers are considered absolutely rigid, except for the points of contact of the cycloidal gear with the pins [5]. The analytical model of this method is presented in fig. 2.a. The pin wheel turns on angle β, when subjected to external torque T. Contact friction is neglected, the reducer is produced without errors, relation between the normal force and the contact displacement is linear.

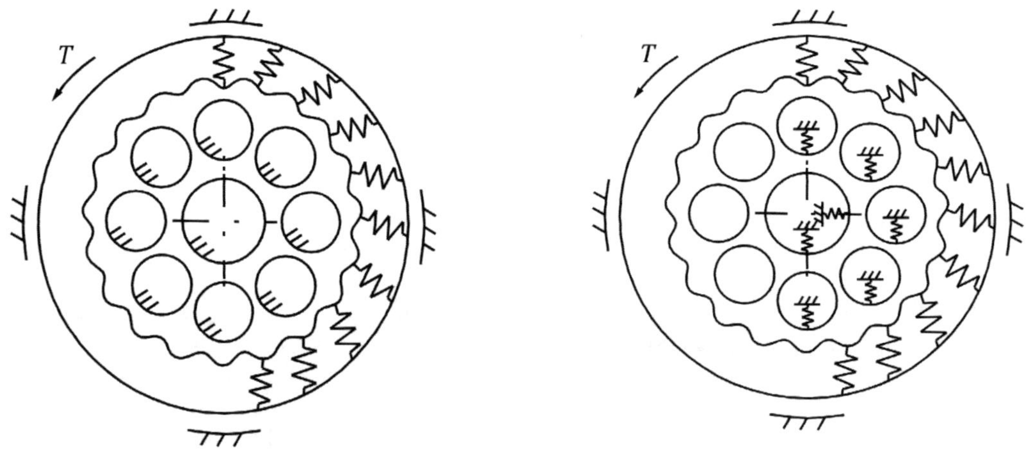

a. existing method b. suggested method

Figure 2. The analytical models of cycloidal gears

Given these assumptions, the strength in the *i*-th pin will be equal to:

$$F_i = r_i\delta_i = r_i\beta(P_i \times N_i), \tag{1}$$

where δ_i – contact displacement of the cycloidal gear with the *i*-th pin [6], r_i – rigidity of contact between cycloidal gear and *i*-th pins, P_i – radius vector of the trochoid profile at the point of contact with the *i*-th pin from the origin of the cycloidal gear, N_i – normal to the trochoid profile at the point of contact with the *i*-th pin (fig. 3).

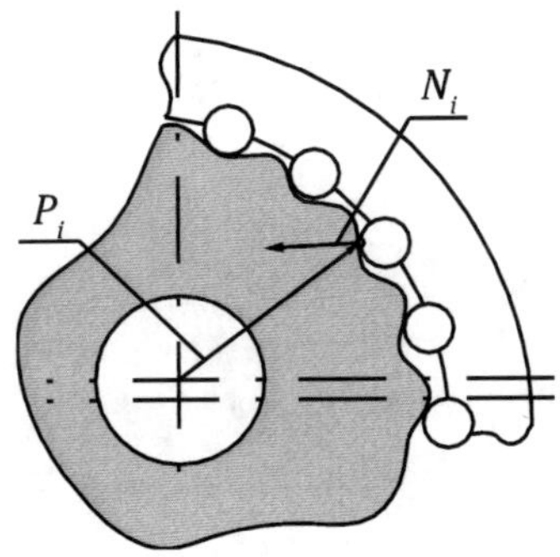

Figure. 3. Graphic representation of the vectors characterizing the geometry of the gearing

Then sum of the moments will be equal to the moment applied to the pin wheel:

$$\sum(P_i \times N_i)^2 \beta = T \tag{2}$$

Following that we substitute [7] the values then the force in the pin will be equal to:

$$F_i = \frac{T}{4(ez_c)^2} \frac{sin(z_c t_i)}{\sqrt{1+\lambda^2-2\lambda cos(z_c t_i)}} \tag{3}$$

The maximum force will act on the pin, which is opposite the pole of the cycloidal gear. This method calculation is valid for the gear of produced with high accuracy, and the distribution of the load among simultaneously operating pins is satisfied.

In order for accounting machining tolerances, the following method was proposed (fig.2.*b*). The cycloidal gear, the pin wheel, the carrier, the disk of the mechanism of parallel cranks and other parts of the reducer are considered absolutely rigid. The points of contact the cycloidal gear with the *i*-th pin and the points of contact the cycloidal gear with the *j*-th finger are considered flexible. Their flexibility is communicated with the relative movement of part's points. Contact friction is neglected. The reducer is produced with errors, which are characterized by the total error vector Δ_i.

Consequently, under the action of external forces, the cycloidal gear moves by the value of u along the x-axis and by the value of v along the y-axis, together this the cycloidal gear is rotated by the angle α and the pin wheel by the angle β. We write these four unknown parameters in the form of a vector:

$$\chi = \{u \quad v \quad \alpha \quad \beta\}^T \tag{4}$$

$\boldsymbol{B}_i$ is denoted the matrix that converts the cycloidal gear and the pin wheel position vector χ into the local displacement vector δ_i:

$$\delta_i = \boldsymbol{B}_i \chi + \Delta_i, \tag{5}$$

F_i is denoted a vector characterizing the loads with which the bearings and contact points act on the cycloidal gear in the direction of local displacements δ_i. We introduce the matrix $\boldsymbol{D}_i$ to convert the vector of local displacements δ_i into the vector F_i,

$$F_i = \boldsymbol{D}_i \delta_i = \boldsymbol{D}_i(\boldsymbol{B}_i \chi + \Delta_i) \tag{6}$$

Denote by $\boldsymbol{R}$ - the vector of external loads acting on the reducer:

$$\boldsymbol{R} = \begin{Bmatrix} F_{cx} \\ F_{cy} \\ 0 \\ T \end{Bmatrix}, \tag{7}$$

where F_{cx}, F_{cy} – projection of centrifugal force, which acts on the cycloidal gear along the axes.

After some calculation we produce:

$$\left[\sum_{i=1}^{n} \boldsymbol{B}_i^T \boldsymbol{D}_i \boldsymbol{B}_i\right]\chi = \boldsymbol{R} - \sum_{i=1}^{n} \boldsymbol{B}_i^T \boldsymbol{D}_i \Delta_i. \tag{8}$$

Solving this system of equations, we obtain the values of global displacements, then we can determine the reactions in the points of contact.

Consider the different types of contact points and the corresponding matrices $\boldsymbol{B}$ and $\boldsymbol{D}$.
For contact points of the cycloidal gear with the p-th pin, the matrix $\boldsymbol{B}$ has the form:

$$\boldsymbol{B_p} = \left\{ \begin{matrix} N_{p_x} \\ N_{p_y} \\ P_{p_x} N_{p_y} - P_{p_y} N_{p_x} \\ e\left(N_{p_y} \cos\varphi - N_{p_x} \sin\varphi\right) + P_{p_x} N_{p_y} - P_{p_y} N_{p_x} \end{matrix} \right\}. \quad (9)$$

The matrix $\boldsymbol{D_p}$ for converting the vector of local displacements δ_p into the vector $\boldsymbol{F_p}$ (if $\delta_p > 0$):

$$\boldsymbol{D_p} = \left| \begin{matrix} r_p, \text{if } \delta \geq 0; \\ 0{,}1 \frac{\text{N}}{\text{mm}}, \text{if } \delta < 0, \end{matrix} \right. \quad (10)$$

where r_p - the rigidity of the contact of the cycloidal gear with the p-th pin.

For contact points of the cycloidal gear with the f-th finger of parallel cranks mechanism, the matrix $\boldsymbol{B}$ has the form:

$$\boldsymbol{B_f} = \left\{ \begin{matrix} N_{f_x} \\ N_{f_y} \\ P_{f_x} N_{f_y} - P_{f_y} N_{f_x} \\ 0 \end{matrix} \right\}^{\text{T}}. \quad (11)$$

The matrix $\boldsymbol{D_f}$ will be equivalent to $\boldsymbol{D_p}$. For contact points of the cycloidal gear with the b- th bearing, the matrix $\boldsymbol{B}$ has the form:

$$\boldsymbol{B_b} = \left\{ \begin{matrix} 1 & 0 & -P_{b_y} & 0 \\ 0 & 1 & P_{b_x} & 0 \end{matrix} \right\}, \quad (12)$$

and matrix $\boldsymbol{D_b}$

$$\boldsymbol{D_b} = \left\{ \begin{matrix} r_b & 0 \\ 0 & r_b \end{matrix} \right\}, \quad (13)$$

where r_b - the rigidity of b-th bearing [8].

The task is solved iteratively, because of the number of working pins is unknown in advance.

3. Case study

The results of the calculation are presented in Figure 4, which shows the strength of constraints effect on the cycloidal gear. The red color indicates the strength of the pins, the green color indicates the fingers, and the blue color indicates the bearings.

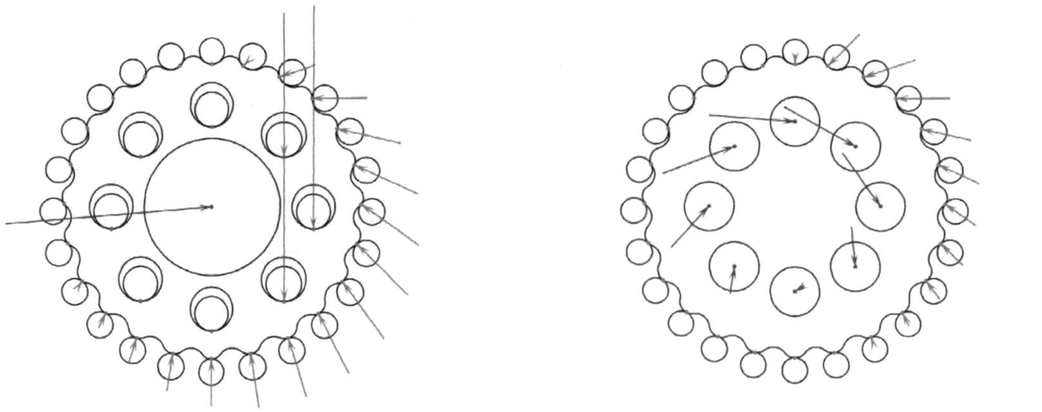

a. forces distribution in KHV without errors b. forces distribution in 2KV without errors

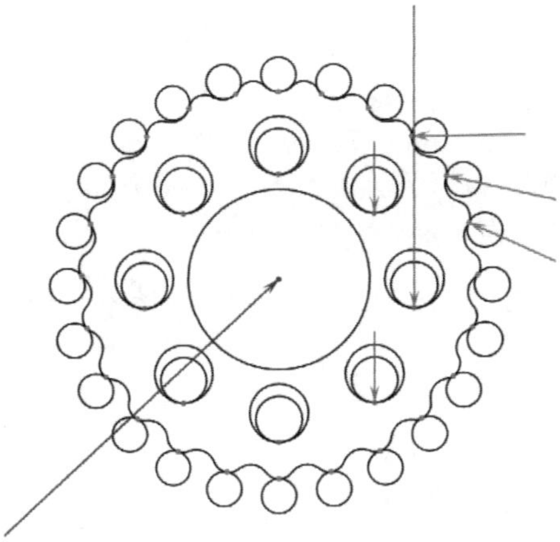

c. forces distribution in KHV with constant error (-0,005 mm) of diameter pins

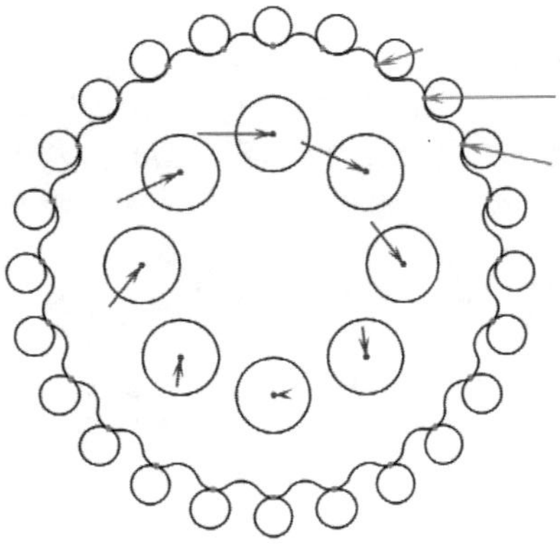

d. forces distribution in 2KV with constant error (-0,005 mm) of diameter pins

e. forces distribution field in KHV with random errors from tolerance (-0.005 mm, 0) of diameter pins

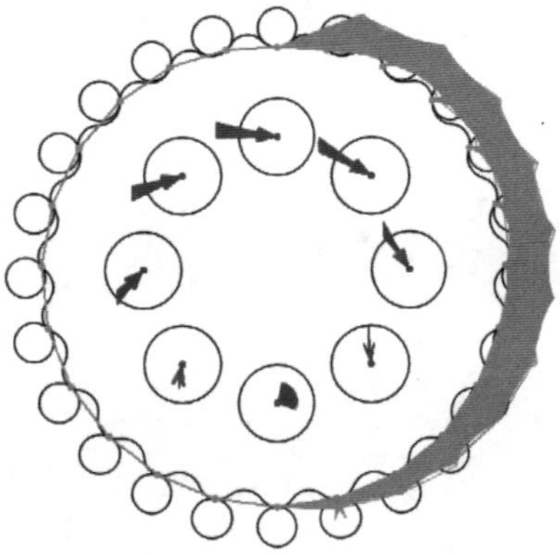

f. forces distribution field in 2KV with random errors from tolerance (-0.005 mm, 0) of diameter pins

Figure 4. The results of the calculation

It will be seen from Fig. 4.*a* and 4.*b*, when calculating a gear produced without any machining errors [9], the distribution of forces of pins corresponds to the traditional method of calculation. If we (Fig. 4.*c*, 4.*d*) add constant machining error (-0,005 mm) of diameter pins, then less number of pins are loaded and the maximum force are increase. In both cases, the KHV scheme is less desired, because of bearing flexibility on the eccentric and the reaction of the mechanism of parallel cranks make a great contribution to forces distribution and the most loaded pin is shifted along of the external moment. However, usually the pins are produced with different accuracy. To calculate, we took random errors from the machining tolerance (Fig. 4.*e*, 4.*f*). Forces distribution fields turned out to be more uniform for reducers KHV and 2KV.

4. Summary

The traditional calculation method gives a good result for 2KV reducer. If we take into account bearings flexibility and the machining tolerances of all parts, then we can obtain a more reliable forces distribution. Accounting for bearings flexibility increases the maximum force in pins by 30%, and accounting for machining errors increases the maximum force in pins twice. In case of cycloidal gear from polyamide the number of working pins increase on 2-3 pieces [10].

IOP Conf. Series: Materials Science and Engineering **842** (2020) 012021 doi:10.1088/1757-899X/842/1/012021

5. References

[1] Zakharov M N, Laryushkin P A, Erastova K G 2018 *Russian Engineering Research.* **38(2)** 72
[2] Laryushkin P A et al 2017 *Russian Engineering Research.* **37(7)** 585
[3] Kudravtsev V N 1966 *Planetarnie peredachi.* (Leningrad: Mashinostroenie) p 308
[4] Ivanov A S et al 2012 *Russian Engineering Research.* **32(11)** 707
[5] Chirkin A V, Ivanov A S, Ermolaev M M *Sovremennoe mashinostroenie. Nauka i obrazovanie.* 462
[6] Ivanov A S, Fomin M V, Ermolaev M M, Chirkin A V 2015 *Russian Engineering Research* **35(3)** 167.
[7] Ivanov A S, Ermolaev, M M, Chirkin A V, Sinitsyna Y V 2015 *MGTU im. N.E.Baumana Moscow Russia*
[8] Ivanov A S, Kuts M S 2018 *Russian Engineering Research.* **38(8)** 573
[9] Ivanov AS, Fomin M V 2017 *Russian Engineering Research.* **37(12)** 1017
[10] Chirkin A 2019 *Spravochnik. Inzhenernyi zhurnal.* 3

IOP Conf. Series: Materials Science and Engineering **842** (2020) 012022 doi:10.1088/1757-899X/842/1/012022

Optimization on Preventive Maintenance and Spare Unit Number for Multi-unit Parallel System

Wenke Gao*, Tianshu Yang, Xufeng Chen and Xin Liang

School of Energy and Power Engineering, Lanzhou university of Technology, Lanzhou, China, 730050

*e-mail: gaowk_best@163.com

Abstract. Preventive maintenance (PM) for multi-unit parallel system with limited production resource is a crucial issue. To find a solution for the issue, this study focused on a multi-unit parallel system with one repairman and some spare units, and developed an optimization model for the system. In the model, PM policy and cold spare unit number were jointly considered. After optimization, an optimal PM policy with PM interval of running and spare units, PM number and cold spare unit number were obtained. Then, a numerical example was analyzed to illustrate the proposed model, and some results were attained based on the sensitive analysis.

Notations

C_f	cost of minimal failure	$h(t)$	failure intensive function
C_p	cost of PM	nd	factor of M
C_r	cost of replacement	n	times of PM for spare units in every PM interval of operating unit
C_d	loss rate of down time	T_p	PM time
T	PM interval of operating-unit	M	operating unit number
T_{st}	PM interval of spare unit	M_s	spare unit number
$\lceil\,\rceil$	integral operation	N_p	PM times of spare unit
$\lfloor\,\rfloor$	rounding operation	N	PM times of operating unit
$g(t)$	the system long-running cost rate		$H(t)=\int_0^t h(u)du$

1. Introduction

A joint consideration of preventive maintenance (PM) for a multi-unit parallel production system with limited production resource is a crucial issue, which widely exists in railway, civilaviation and production fields. The configuration of production resource, such as spare unit, repairman or maintenance team, is a main part in production process. A rich configuration can cause waste of resource, while a poor one may result in outage. Therefore, how to configure spare unit for the multi-unit production system with one repairman based on a minimal running cost rate is an exceedingly valuable research issue. To resolve this issue, maintenance, operating unit number, repairmen and production should be jointly considered in the resource configuration and maintenance decision-making[1-3].

Research on resource configuration, Ni et al, [4] pointed out that production is often interrupted by pre-scheduled PM without considering the throughput target. PM tasks are not cost-effectively conducted, and investigated the extra hidden opportunities for PMs during production time without violating the system throughput requirement. Liao et al.[5] oriented a mono-machine-based optimization model of production scheduling and PM under group production is proposed. Xiao et al.[6] proposed a joint optimization model to minimize the total cost including production cost, PM cost, minimal repair cost for unexpected failures and tardiness cost. Hidayanto et al.[7]take the salt crusher machine as object, proposed four PM plans based on the reliability analysis, compared the cost, reliability and production of these PM plans, and obtained optimal PM plans in the cost, reliability and production. Some research regarded that grouping replacement or maintenance for all units of multi-unit parallel production system can reduce setup cost of group[8]. Pargar et al.[9] developed an integrated optimization method to schedule PM and renewal projects by grouping them and simultaneously finding the optimal balance between them. Do et al.[10] proposed a dynamic maintenance grouping approach for multi-unit system, and the approach can provide a minimum number of repairman to ensure the maintenance plan. Bai et al.[11] introduced grouping maintenance strategy for optimizing the compound maintenance intervals of multi-component system, and proposed the optimization steps and methods. However, grouping maintenance policy can result in outage to production system, and the outage is not agreed in some fields. Zhang et al.[12] developed a PM policy which jointly considered the PM policy and the operating unit number for a multi-unit parallel production system. Whereas the number of a parallel running unit is a fixed value for most multi-unit parallel system, and usually configured by engineering experience. Therefore, it is necessary to optimize the number of cold spare unit under a fixed running unit system.

2. Maintenance and assumption

In this section, a multi-unit parallel production system with M operating units, M_s spare units and one repairman will be considered. In this system, the operating and spare units arise out of the same type. In accordance with the schedule plan, the sequentially PM activities of operating units are performed at $kT(k=1,2,\ldots,N-1)$ and a replacement is carried out at NT. As for the spare units, keeping the same running time with operating unit is hard to achieve, so if the accumulated operating time of spare unit T_{st} is within $[\alpha_1 T, \alpha_2 T]$ ($\alpha_1<1$, $\alpha_2>1$), it will be preventively maintained. In order to ensure the normal implementation of the schedule plan, one of the spare units performs operating when one of the operating units executes PM activity. The repairman can maintain only one unit at the same time and just for PM activity.

Take M=6, N=4 and M_s=2 as an example to introduce the operation and maintenance process. The parameter n means the times of PM for spare units in every PM interval of operating units. As for this system, let the parameter $n=1, 1\frac{2}{4}, 1\frac{1}{5}, 2$ and 3. Then, the process diagram respectively showed as subplot (a) to (e) of Fig.1.

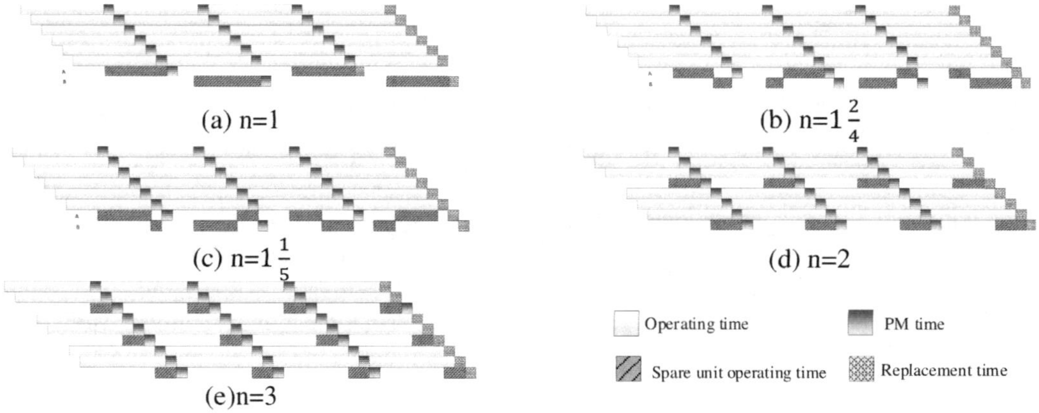

Figure 1. The operating process of system.

To convenience for modeling, the following assumptions are proposed firstly,

1) All failures can be discovered and repaired immediately, and the time for repair and replacement are ignored, while the time of PM is considered.

2) The hazard rate function of a new unit is $h(t)$, which is monotonically increasing with respect to t and each PM can make it return to 0. PM can change the hazard rate of unit following $h_i(t)=a_i h(t)$ (i=1,2,…,N) before the ith PM, where $1=a_1<a_2<\ldots<a_N$. The unit can be restored to "as good as new" state after repair.
3) The cost of minimal failure C_f is independent of the time of the failures and the total sever of the system. The cost of PM C_p is irrelevant to the total operational time of the system. The loss rate of outage time of spare unit is C_d. After the N-1th PM, the operating unit is replaced and the cost is a constant C_r, where $C_r>C_f$.
4) The PM activity of all units can be finished in the PM time. PM interval of each operating unit is T while that of the spare unit is T_{st} which is within $[\alpha_1 T, \alpha_2 T]$, where $\alpha_1\leq 1$, $\alpha_2\geq 1$.

3.Modeling and optimization

In this section, the long-term running cost rate and optimization model of system will be illustrated in detail.

3.1 Modeling

As for the multi-unit production system, a renewal cycle is defined as the time from when the system starts running until all units finish the first replacement, or the interval between two successive replacements of all units which consist of M operating units and M_s spare units.

In the PM model, three actions are comprehensively considered: minor repairs, PMs and replacements. Minor repairs can restore the minor failures and the replace cycle is renewal process. Let Y denote the length of a replace cycle, C means the total cost of renewal cycle and $g(T, n, T_{st}, N, M_s)$ denote the long-term running cost rate of system within the interval $(0, t)$. Thus, (Y, C) constitutes a renewal reward process. In accordance with renewal theory, the function of total s-expected long-term running cost rate of renewal cycle can be described as the Eq.(1).

$$g(T,n,T_{st},N,M_s)=\frac{E[C]}{E[Y]}=$$
$$\frac{\left\{\begin{matrix} C_f\left(\begin{matrix} M\sum_1^N H_i(T)+M_s\sum_1^{\lfloor Nn/M_s\rfloor}H_i(T_{st})+\lfloor Nn-M_s\lfloor Nn/M_s\rfloor\rfloor H_{\lceil Nn/M_s\rceil}(T_{st})+ \\ H_{\lceil Nn/M_s\rceil}\left(\left(Nn-\lfloor Nn/M_s\rfloor-\lfloor Nn-\lfloor Nn/M_s\rfloor\rfloor\right)T_{st}\right)\end{matrix}\right) \\ +C_p\left(M(N-1)+N_p\right)+C_r(M+M_s)+C_d\left(M_sN\left(T+T_p\right)-MNT_p-N_pT_p\right)\end{matrix}\right\}}{NM(T+T_p)} \tag{1}$$

$$N_p=\begin{cases}0 & Nn\leq M_s \\ \lceil Nn\rceil-M_s & other\end{cases} \tag{2}$$

$$T\begin{cases}\geq (M-1)T_p & N_p\leq 0 \\ \geq MT_p & 0<N_p\leq N-1 \\ \geq\left(M+\left\lceil\frac{N_p}{N-1}\right\rceil-1\right)T_p & other\end{cases} \tag{3}$$

$$T_{st}=MT_p/n \tag{4}$$

$$\alpha_1 T\leq T_{st}\leq\alpha_2 T\Leftrightarrow\frac{T_{st}}{\alpha_2}\leq T\leq\frac{T_{st}}{\alpha_1} \tag{5}$$

In the above formula, Np is the number of PM for spare unit and T_{st} is the interval between the adjacent PMs for spare unit. In addition, n is the times of PM for standby units in every PM interval, and the relationship among n, M_s, N and M is shown in Table1.

Table 1. The relationship among n, M_s, N and M.

M_s	n										
1	$1/k$	1	--	nd_1	--	nd_2	--	…	nd_i	--	M
2	$1/k$	1	$M/(M-k)$	nd_1	--	nd_2	--	…	nd_i	--	M
3	$1/k$	1	$M/(M-k)$	nd_1	$2M/(M-k)$	nd_2	--	…	nd_i	--	M
…	$1/k$	1	$M/(M-k)$	nd_1	$2M/(M-k)$	nd_2	…	…	nd_i	--	M
ms	$1/k$	1	$M/(M-k)$	nd_1	$2M/(M-k)$	nd_2	…	$(ms-1)M/(M-k)$	nd_i	--	M
	$(k=1,\ldots,N)$		$k=1,\ldots,M-1$		$k=1,\ldots,M/nd_1$		…	$k=1,\ldots,M/nd_i-1$			

3.2 Optimization

To obtain the optimal PM policy and spare unit number, the system long-term running cost rate is to be minimized. To find the optimal results, the following theorems are introduced first.

Theorem 1: Under a given M_s and N. If the hazard rate function $h(t)$ is a monotonous increasing function of t, the optimal T^* and T_{st}^*will be obtained.

To jointly optimize PM and the number of spare unit, we need to minimize the long-term running cost rate $g(T, n, T_{st}, N, M_s)$.Therefore, let$dg(T, n, T_{st}, N, M_s)/dT = 0$ and get Eq.(6).

$$\sum_{i=1}^{n}(T+T_p)h_i(T) - H_i(T) = \frac{C_f\begin{pmatrix} M_s\sum_1^{\lfloor Nn/M_s\rfloor} H_i(T_{st}) + \lfloor Nn - M_s\lfloor Nn/M_s\rfloor\rfloor H_{\lceil Nn/M_s\rceil}(T_{st}) + \\ H_{\lceil Nn/M_s\rceil}\left((Nn - \lfloor Nn/M_s\rfloor - \lfloor Nn-\lfloor Nn/M_s\rfloor\rfloor)T_{st}\right)\end{pmatrix} + C_p\left(M(N-1)+N_p\right) + C_r(M+M_s) - C_d\left(M_sNT_p + MNT_p + N_pT_p\right)}{C_fM} \quad (6)$$

In accordance with the assumption (2) and the given M_s and N, the above equation existsa unique solution. Thus, the optimal T^*could be obtained in accordance with the theorem 1 in reference[13].According to Eq.4 and Table 1, the relationship between T_{st} and T and T_{st} can be got because n can be obtained when M_s, M and N are given. Thus, the optimal T^* and T_{st}^* can be attained because the $h(t)$ is a monotonous increasing function of t. The detailed proof is negligible.

Theorem 2: Under a given M_s and T. If the hazard rate function $h(t)$ is a monotonous increasing function of t, the optimal N^* will be obtained.

According to the theorem 2 in the reference[13],under a given M_s and T, the optimal N^*can be obtained since $h(t)$ is a monotonous increasing function in t . The detailed proof is negligible.

Theorem 3: Under a given N and T. If the hazard rate function $h(t)$ is a monotonous increasing function of t, the optimal M_s^* will be obtained.

The proof is the same as theorem 2 in the reference [13]. The detailed proof is negligible.

4. Numerical case study

4.1. Numerical example

To illustrate the proposed model, a Weibull distributed example will be provided to demonstrate the character of the model in this paper. Hence, that the lifetime of each unit obeys Weibull distribution and the hazard rate function of new unit as below.

$$h(t) = 0.042\left(\frac{t}{50}\right)^{1.1} \quad (7)$$

Now, further assume that M=12, T_p=4.5, a_i=4i/(3i+1), C_f=2.5, C_p=2, C_r=3, C_d=0.01, α_1=0.5 and α_2=1.5.Then, let the initial values of M_s and N be 1,The step size of iteration for M_s and N respectively are δM_s and δN, and the value of δM_s and δN are 1.An optimal algorithm shown as Fig.2 is used to compute the optimal results.

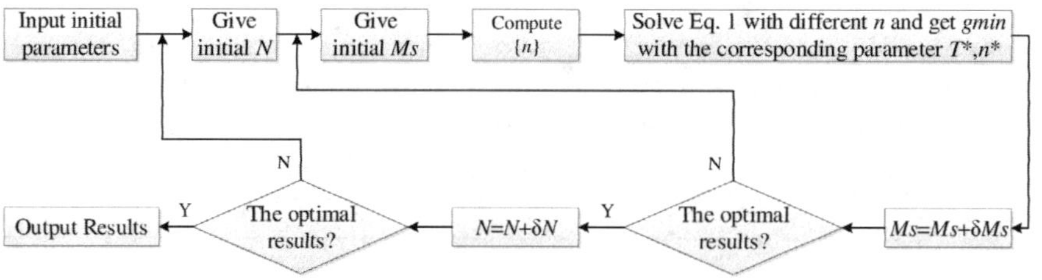

Figure 2. The optimal algorithm.

With the above algorithm, the optimal results of N, M_s, n, T and T_{st} are obtained. The optimal results presented in Table2.

Table2. Optimal results.

N	2	3	4	5	6	7
M_s	1	2	2	2	2	2
n	1	1.5	1.5	1.3333	1.2	1.2
T	54	49.5	49.5	49.5	49.5	49.5
T_{st}	54	36	36	40.5	45	45
g	0.1046	0.1036	0.1029	***0.10281***	0.10284	0.1031

From the above, the optimal $N^*=5$, $M_s^*=2$, $T^*=49.5$ and $T_{st}^*=M^*\times Tp/n^*=40.5$ can be obtained, and the minimal long-term running cost rate of system is 0.10281. That is to say, one repairman can serve a parallel system with 12 operating units and 2 spare units, and the system could be taken preventive maintenance each 5 PM intervals and taken replacement action at the end of the last interval. In addition, the PM interval of operating units and spare units are 49.5 and 40.5 respectively.

4.2 Sensitive analysis

Due to the parameters T_p and M are crucial for the proposed model. And thus, the influence of T_p and M will be analyzed in this section.

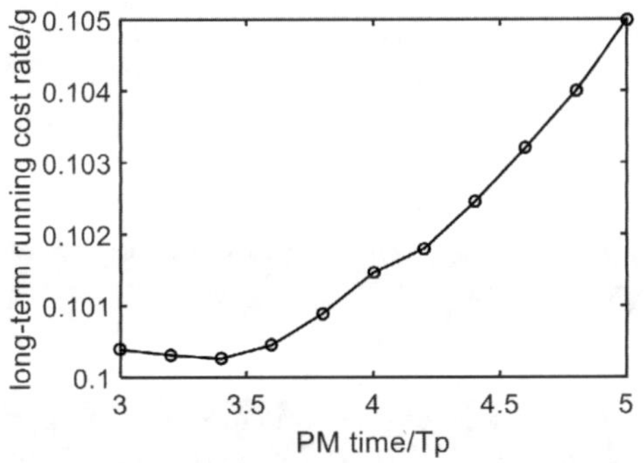

Figure 3. Plot of g-T_p.

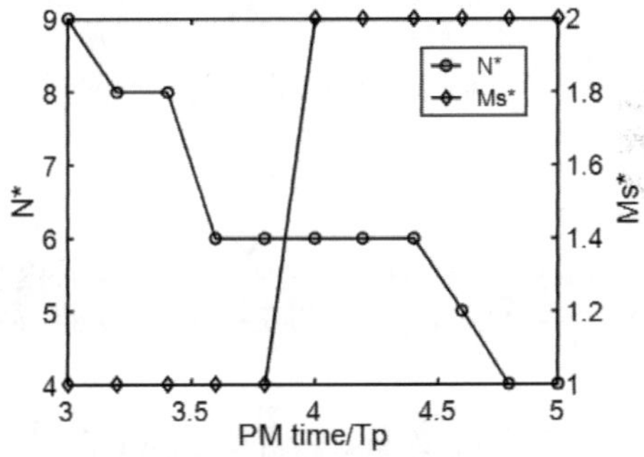

Figure 4. Plot of N^*/M_s^*-T_p.

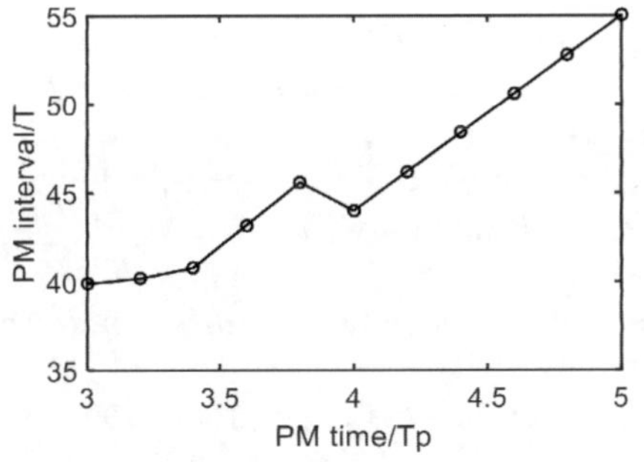

Figure 5. Plot of T-T_p.

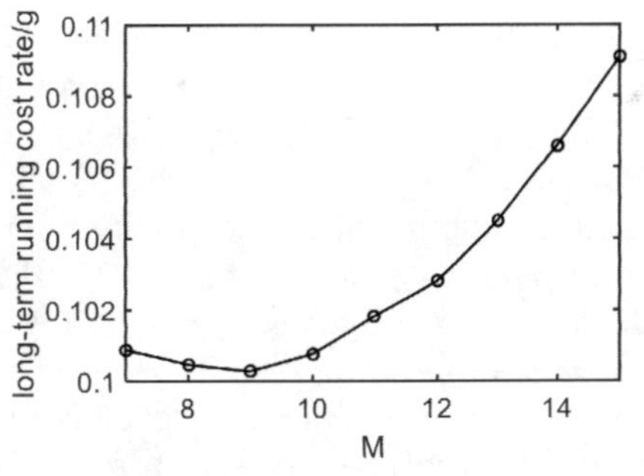

Figure 6. Plot of g-M.

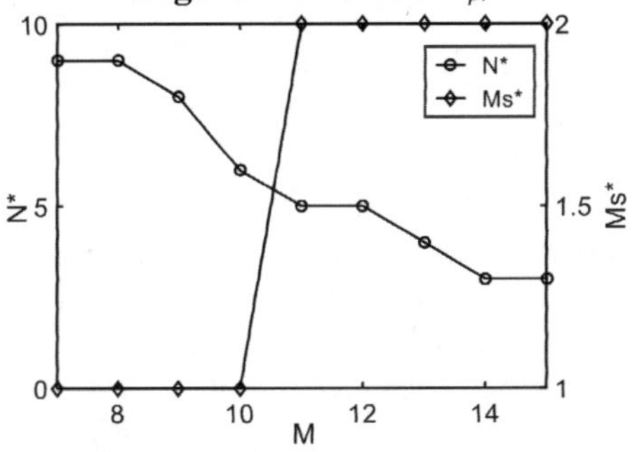

Figure 7. Plot of N^*/M_s^*-M.

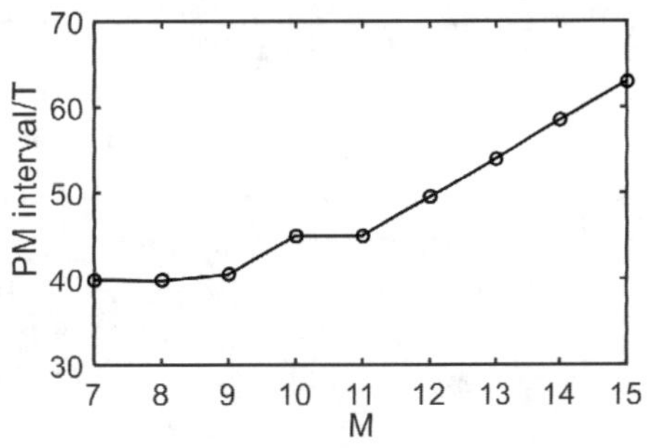

Figure 8. Plot of T-M.

Fig. 3 shows that the optimal value of T_p is existent to minimize the long-term running cost rate. The optimal value of T_p is 3.4 and *gmin*=0.1026. It means that the enterprises could decrease the cost by keeping the PM time at optimal value.Fig.4 indicates that the value of N is decreasing and the value of M_s is increasing with the increase of T_p. In addition, the change of N is clearer than M_s.Fig.5 displays that the PM interval is increasing in the interval [3, 3.8] and [4, 5], the turning point is the same as the curve of M_s in Fig.4.

Fig.6 exhibits that the long-term running cost rate has a minimum value 0.10027 when the value of M is 9. That is to say, the enterprises can adjust the parameters according to the change of the number of operating units to keep the minimum long-term running cost rate.Fig.7 shows that the value of N is decreasing with the increase of T_p while the value of M_s is increasing. The N has more significant effect than *Ms*.Fig.8 indicates that the value of T has a trend of increase with the growth of M. In addition, the turning point is the same as the curve of M_s in Fig.7.

5. Conclusion

In this paper, the joint optimization for multi-unit parallel system with one repairman and several spare units is studied. According to the renewal theory and renewal reward process, the function of long-term running cost rate for the proposed model is established. The optimization objective is minimizing the long-term running cost rate. A numerical example is used to prove the existence and uniqueness of the optimal solution. Afterwards, the sensitive analysis of two significant parameters is given to demonstrate the influence on the optimal solution. The results of sensitive analysis show that the PM time and the number of operating units both have the optimal solution that minimizes the long-term running cost rate of system. These results exhibit that the enterprises could decrease the long-term running cost rate by adjusting the value of two parameters.

Acknowledgment

The research work is also supported jointly by the National Natural Science Foundation of China (grant number 71561016), Postdoctoral Science Foundation (grant number 2017M613297XBand 2019T120964).

Reference

[1] Dellagi S., Chelbi A., and Trabelsi W., "Joint integrated production-maintenance policy with production plan smoothing through production rate control," *Journal of Manufacturing Systems,* vol. 42, pp. 262-270, 2017.

[2] Duan C., Deng C., Gharaei A., Wu J., and Wang B., "Selective maintenance scheduling under stochastic maintenance quality with multiple maintenance actions," *International Journal of Production Research*, pp. 1-19, 2018.

[3] de Jonge B., and Scarf P. A., "A review on maintenance optimization," *European Journal of Operational Research*, 2019.

[4] Ni J., Gu X., and Jin X., "Preventive maintenance opportunities for large production systems," *CIRP Annals,* vol. 64, no. 1, pp. 447-450, 2015.

[5] Liao W., Zhang X., and Jiang M., "An optimization model integrated production scheduling and preventive maintenance for group production." pp. 936-940.

[6] Xiao L., Song S. L., Chen X. H., and Coit D. W., "Joint optimization of production scheduling and machine group preventive maintenance," *Reliability Engineering & System Safety,* vol. 146, pp. 68-78, Feb, 2016.

[7] Hidayanto T. E., Nugroho H. S., and Ardi M. M., "Reliability analysis for preventive maintenance of salt crusher machine," *International Journal of Mechanical Engineering and Robotics Research,* vol. 8, pp. 297-303, 01/01, 2019.

[8] Chiu C.-W., Chang W.-L., and Yeh R.-H., "Group Replacement Policies for Repairable N-Component Parallel Systems," *Engineering Asset Management 2016*, pp. 25-38: Springer, 2018.

[9] Pargar F., Kauppila O., and Kujala J., "Integrated scheduling of preventive maintenance and renewal projects for multi-unit systems with grouping and balancing," *Computers & Industrial Engineering,* vol. 110, pp. 43-58, 2017.

IOP Conf. Series: Materials Science and Engineering **842** (2020) 012022 doi:10.1088/1757-899X/842/1/012022

[10] Do P., Vu H. C., Barros A., and Bérenguer C., "Maintenance grouping for multi-component systems with availability constraints and limited maintenance teams," *Reliability Engineering & System Safety,* vol. 142, pp. 56-67, 2015.

[11] BAI Y., JIA X., and CHENG Z., "Group optimization models for multi-component system compound maintenance tasks," *Maintenance and Reliability,* vol. 49, no. 1, pp. 42-47, 2011.

[12] Zhang W., Gao W., Wang Z., Li B., and Liu X., "Joint Optimization on Preventive Maintenance and Operating Unit Number for Multi-unit Parallel Production System with One Repairman and Spare Unit." pp. 833-838.

[13] Gao W., Zhang Z., Ji H., Zhou Y., and Liu Q., "Optimal Quasi-Periodic Preventive Maintenance Policies for Repairable System with Dynamic Maintenance Plan," *Eksploatacja iniezawodnosc-Maintenance and Reliability,* vol. 17, no. 3, pp. 389-397, 2015.

IOP Conf. Series: Materials Science and Engineering **842** (2020) 012023 doi:10.1088/1757-899X/842/1/012023

Job Shop Layout Improvement with Simulation Technique: A Case of Metal Sheet Production

Wimalin Laosiritaworn*, Nuchjira Pinchai, Panitan Jarupuncharadsri

Department of Industrial Engineering, Faculty of Engineering, Chiang Mai University, Chiang Mai, Thailand

E-mail: wimalin@hotmail.com

Abstract. This paper investigates the improvement of job shop layout with simulation technique. A case study of Siam GN metal sheet, the manufacturer of customized metal sheet products was used. The objective is to increase production efficiency and reduce materials handling distance and time. Data regarding product and machine type were collected. CORELAP (Computerized Relationship Layout Planning) and Group technology based on ROC (rank order clustering) algorithm were applied to design the improved layouts. ARENA software was used to simulate the new layout from both techniques. The simulation show that CORELAP layout reduce production time by 17.12% while group technology layout reduces materials handling time by 11.41%. CORELAP layout therefore, was chosen for implementation. Simulation techniques allows user to experiment with different scenario hence selecting the most appropriate one without interfering with the actual process.

1. Introduction

Job shop is a small manufacturing process that produce customized products in small batches to serve numbers of customer in the same time. Typical layout used in job shop is the 'process layout' in which similar activities or machines are located together in the same department. The advantage of this type of layout including high flexibility, difficult to disrupt and easily supervised (1). In this kind of layout, products move through departments with different paths according to their production route. However, some departments might have high movement frequency; these departments should be located close to each other in order to minimize material handling.

Typically, layout problems are related to the location of machines and departments in a factory. Most of these problems are NP hard (2) To find the layout that minimize material handling with optimization algorithm is very difficult due to the nature of the problem in job shop that involve large number of product, which follows different path through the factory. Various methods have been used to design layout that minimize material handling distance. This research utilizes simulation approach, which the simulation of actual production process was developed as computer model. This model was used to test different layouts in order to find the layout with minimum material handling distance. Simulation model helps the test various layout scenario without interfering with the actual process.

To demonstrate the use of simulation techniques, a case study of a manufacturer of metal sheet production at Siam GN metal sheet was used. The company is located in Lamphun province in the northern part of Thailand. The company has 4 main product groups as shown in Figure 1; namely metal roof sheet, roof flashing, galvanized beam and sheet metal fence panel.

a) Metal roof sheet b) roof flashing c) galvanized beam d) metal fence panel

Figure 1. Products of Siam GN metal sheet.

2. Literature review

The research applied CORELAP and Group technology to design new layout, the related concept regarding both techniques are briefly describe in this section.

2.1. CORELAP

This algorithm was introduced in 1967 by Roberr C. Lee and Moore (3). CORELAP algorithm is based on total closeness rating (TCR). It develops layouts by location rectangular-shaped departments. The TCR rating is the sum weighted rating between the new department and its neighbor in the layout (4).TCR for each department can be calculated using the following equation

$$TCR_i = \sum_{j=1}^{m} r_{ij} \quad (1)$$

where, m = number of departments

2.2. Group technology

Group Technology is a method that identifies the similar attribute of product design and manufacturing process in order to identify machine groups and part families to crate manufacturing cell with minimum moves into other cells (5). The research employ rank order clustering (ROC) method which as following algorithm (6)

Step 1: Create n*m matric b_{ij}(binary number for part and machine), where n in parts and m is machines

Step 2: For each row of i, compute $\sum_{i=1}^{m} b_{ij} * 2^{m-j}$

Step 3: Rearrange the rows in descending order based on the computed numbers

Step 4: For each row of j, compute $\sum_{j=1}^{n} b_{ij} * 2^{n-i}$

Step 5: Rearrange the columns in descending order based on the computer numbers

Step 6: Repeat step 1 until there no change is observed in step 3 and 5

Step 7: Stop

3. Research methodology

The research methodology started with data collection, and then two layouts were designed based on CORELAP and ROC. Finally, both layouts were compared using simulation method. The detail methodology is described as follows.

3.1. Data collection

In this stage product related data including type, size, shape and demand volume were collected. Next, multi product process chart was used to summarize routing of each product. From to chart was used to collect both distance and frequency of materials handling between machines. Distance was calculated using rectilinear formula. For instance the distance between machines that has co-ordinate of (x_i, y_i) and (x_j, y_j) can be calculated from $|x_i - x_j| + |y_i - y_j|$

3.2. CORELAP layout design

Composite movement frequency between machines were calculated from data in from-to chart. The frequency obtained is coded in to A, E, I, O, U and X where A represent highest relationship and X is undesirable relationship. TCR is calculated by substitute A = 10000, E = 1000, I = 100, O = 10, U = 0 and X = -10000. Machines are then assigned according to CORELAP algorithm.

3.3. ROC layout design

Part-machine matrix was construct where column represent the 14 different products and row represent the 23 machines in the shop. If the production of the specific product require that specific machine, the value in that correspond cell become 1, otherwise 0. Then layout was developed according to ROC algorithm described in section 2.2.

3.4. Simulation

Arena software was used to develop the simulation model of the factory. The first model is the current layout, which represents the baseline of the analysis. This model was verified by comparing number of finished product obtained from the model with number of input. The criteria of acceptance is that the different should be less than 5 lots. The model was also validated by comparing average time of actual process with average time of model process and the different should be less than 5%. After the model pass the verified and verification processes, the model of CORELAP layout and ROC layout was developed and the results from the 3 models were compared.

4. Results and Discussion

4.1. Basic information regarding the case study company layout

The majority of product manufactured in the case study company is metal roof sheet (74%), roof flashing (15%), sheet metal fence panel (8%) and then galvanized beam (3%). Each product category can be divided in to sub-category. For example, metal roof sheet has 5 different shapes and sheet metal fences has 2 different sizes. There are total of 15 sub-categories.

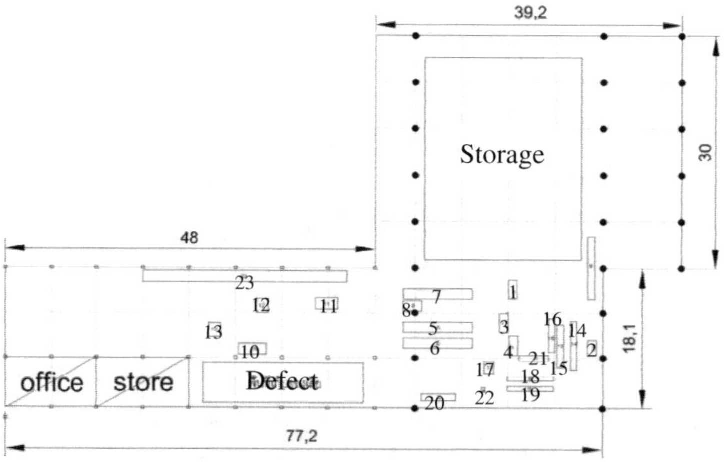

Figure 2. Products of Siam GN metal sheet.

Multi product process chart is shown in Table 1 with the total of 23 machines and 15 products sub-category. As can be seen each product follows different path through out factory.

4.2. New layout design

The two new layout design are shown in Figure 3. CORELAP block diagram shown in Figure 3a) was developed by firstly place machine 5 in the matrix, then machine 3 and 17 which have close relation ship with machine 5 was located close to machine 5. The process repeat until the final block design was obtained. The block diagram was expanded to fit actual space by taking in to account of the space requirement of each machine.

Table 1. Multi product process chart.

		1	2	3	4	5	6	7	8	9	10	11	12	13	14	15
1	Decolier 1			1												
2	Decolier 2															1
3	Decolier 3	1			1		1	1	1	1	1	1	1			
4	Decolier 4		1			1								1	1	
5	Roof forming 750	2					2	2	2	2	2	2	2			
6	Roof forming 760		2			2								2	2	
7	Roof forming (Spain)			2												
8	Roof forming				2											
9	C beam forming															2
10	Roll forming with PE 750	3	3													
11	Roll forming with PE			3												
12	Crimping machine 750	4														
13	Crimping machine 760		4													
14	Barge rolling								4							
15	Ridge rolling										4					
16	Louvres rolling						4									
17	Cutting machine						3	3	3	3	3	3	3		3	
18	Small fence forming													4		
19	Large fence forming														4	
20	Folding machine							4		4		4		3		
21	Folding machine															
22	Crimping machine						5	5	5	5	5	5	4			
23	Foam injection					3										

ROC layout design are shown in Figure 3b). This layout was developed by firstly applying ROC algorithm to group products into 2 groups. Then the machine used to manufacture those products were group accordingly and two separate cells was designed in the new layout.

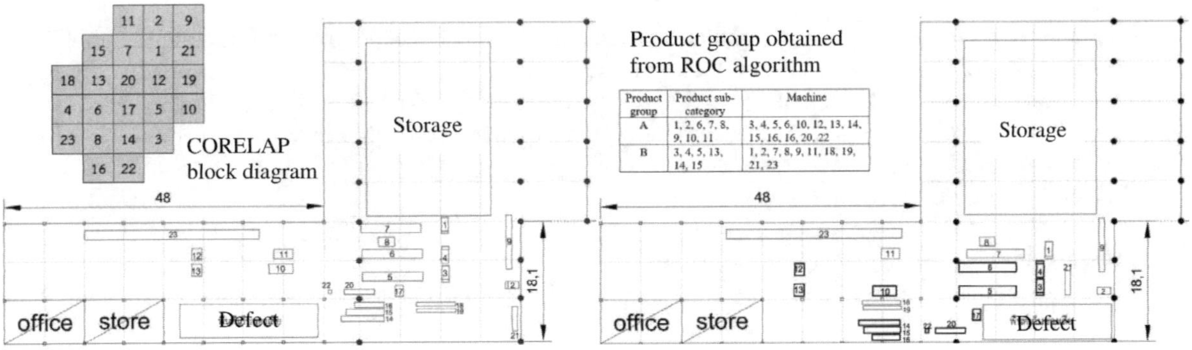

Figure 3. New layout design.

4.3. Simulation models

Arena software was used to simulate actual production process. 5 models were developed for roof type 750, roof type 760, roof flashing and fence, roof spain type and C-beam. An example of a model of roof flashing and fence is given in Figure 4. The simulation started with create module. In this step, entity name and quantity were specified. Then the entity type was assigned with assign module. Next, process module was used to simulate production process with production time and distribution identified from input analyzer. Decide module was used next to decide which entity enter which machine.

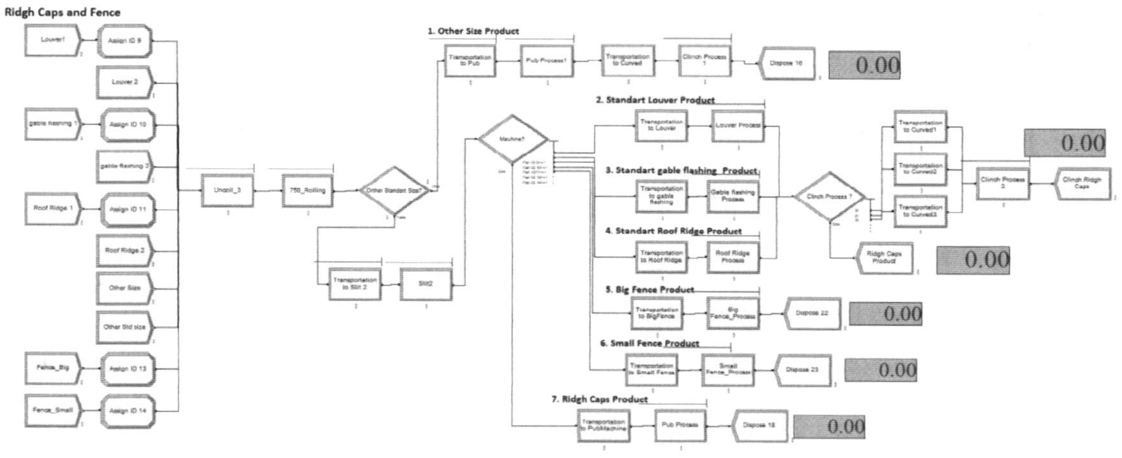

Figure 4. Simulation model for roof flashing and metal fence panel.

Table 2 compare current layout with the two new layout. In terms of material handling distance, CORELAP layout reduce 13.66% of distance in comparison with current layout while ROC layout only reduce distance by 4.37%. Similarly, total production time was reduced more with CORELAP layout at 17.12%. Therefore, CORELAP layout is superior in both criteria and was chosen for actual implementation.

Table 2. Models comparison.

	Material handling distance	% decrease in distance	Total production time (min)	% decrease in time
Current layout	355.70	n/a	2110.97	n/a
CORELAP layout	307.11	13.66%	1749.64	17.12%
ROC layout	340.14	4.37%	1870.058	11.41%

5. Conclusion

This paper presents the application of simulation model in order to select the best layout which minimize materials handling distance and total production time. The research start with collecting data regarding product, production routing and materials handling frequency. Two layout was designed based on CORELAP and ROC algorithm. Arena simulation was used to simulate current layout and the two improved layout. CORELAP layout was found to be superior to ROC layout in both material handling distance and total production time.

References

[1] Slack N, Brandon-Jones A, Johnston R. Operations Management. seventh ed: Pearson Education Limited; 2013.

[2] Drira A, Pierreval H, Hajri-Gabouj S. Facility layout problems: A survey. Annual Reviews in Control. 2007;31(2):255-67.

[3] Sembiring A, Budiman I, Mardhatillah A, Tarigan U, Jawira A. An application of corelap algoritm to improve the utilization space of the classroom. Journal of Physics: Conference Series. 2018;1007:012026.

[4] Khan AJ, Tidke D, Scholar M. Designing facilities layout for small and medium enterprises. International journal of Engineering research and General science. 2013;1(2).

[5] Alhourani F. Clustering algorithm for solving group technology problem with multiple process routings. Computers & Industrial Engineering. 2013;66(4):781-90.

[6] Amruthnath N, Gupta T. Modified Rank Order Clustering Algorithm Approach by Including Manufacturing Data. IFAC-PapersOnLine. 2016;49(5):138-42.

IOP Conf. Series: Materials Science and Engineering **842** (2020) 012024 doi:10.1088/1757-899X/842/1/012024

Predictive Manufacturing: Subtractive and Additive

Steven Y. Liang[1,*], Yixuan Feng[1], Jinqiang Ning[1]

Woodruff School of Mechanical Engineering, Georgia Institute of Technology, Atlanta, GA, 30332
E-mail: steven.liang@me.gatech.edu

Abstract: Manufacturing is the key to today's industrial competitiveness, and it is broadly classified into two categories, subtractive and additive. In current study, the ability to predictively model manufacturing performance attributes in both categories is introduced. In subtractive manufacturing, modeling of laser-assisted and ultrasonic vibration-assisted milling are presented. In laser-assisted milling, the laser preheating temperature field is predicted, and the dynamic recrystallization as well as grain growth triggered under high temperature is considered, which enhances the accuracy of force and residual stress prediction. In ultrasonic vibration-assisted milling, the intermittent effect is considered through tool-workpiece separation criteria. And the force reduction in ultrasonic vibration-assisted milling is accurately predicted. In additive manufacturing, laser-assisted metal additive manufacturing is introduced. And the predictive modeling of temperature field in powder bed metal additive manufacturing is presented. The model considers heat transfer boundary including heat loss from convection and radiation at the part boundary. Through the comparison between measured and calculated molten pool dimensions, the model is proven to have high computational efficiency and high prediction accuracy.

1. Introduction

1.1. Subtractive manufacturing

In subtractive manufacturing, advanced machining technologies have been developed for years in order to obtain higher material removal rate, better surface finish, and better fatigue resistance, while keeping tool wear rate and cutting force at a low level. In addition, the use of these technologies could enhance the machinability of some difficult-to-machine materials such as Inconel 718 and Ti6Al4V, which typically have high strength and cause large cutting forces, high cutting temperature, severe tool wear, and bad surface finish in conventional machining process. Laser-assisted and ultrasonic vibration-assisted milling are two of the advanced technologies applied in subtractive manufacturing. The former reduces cutting force by softening the workpiece through laser preheating, while the latter has the same effect through discontinuous machining process under ultrasonic vibration of the tool. However, with the use of advanced technologies such as laser and ultrasonic vibration, more challenges appear for the process control. For example, under high cutting speed, the use of laser could lead to higher flank wear or even tool breakage [0] comparing with conventional milling, when inappropriate combination of cutting and laser parameters are used. The ultrasonic vibration could also speed up the tool wear rate and further lead to larger surface roughness than conventional milling [0]. Therefore, predictive modeling is important to calculate force, temperature, residual stress, tool wear, and surface roughness in laser-assisted and ultrasonic vibration-assisted milling process.

Several predictive models have been proposed so far. Pan *et al.* [0] predicted force in laser-assisted milling of Inconel 718. The oblique cutting is transferred to equivalent orthogonal cutting at each rotation angle. Then based on mechanics of machining, the forces in cutting and radial directions are calculated. The axial force is calculated according to tool geometry. However, the preheating temperature is calculated through numerical simulation, and the microstructure evolution is not considered. Tian *et al.* [0] proposed thermal modeling of laser-assisted milling to predict temperature field by determining spatial distribution of the laser energy and heat transfer equations including conduction, convection, and radiation. Although good agreement was found, the validation heating tests were conducted without material removal. Therefore, only the laser preheating temperature was predicted without machining induced temperature rise. Ulutan *et al.* [0] proposed an analytical model for residual stress, in which the thermal and mechanical loading were computed through elastic-plastic model and relaxation process. Huang *et al.* [0-0] proposed the tool wear prediction model in hard turning of Cubic Boron Nitride cutters, which considered abrasion, adhesion, and diffusion. These models are only in orthogonal cutting or turning process without laser effect. Hao *et al.* [0] developed surface roughness prediction model of ball end milling cutter. The geometrical model of the tool was built by considering tool path, cutting edge, and tool axis. The model was then corrected under the deformation caused by milling force. But the response of workpiece was not included.

Verma *et al.* [0] predicted static machining force in axial ultrasonic vibration-assisted milling. The instantaneous chip thickness is calculated based on the frequency and maximum velocity of vibration, and the mean cutting force for the specific tool angular position is derived from contact ratio and shear flow stress based on Johnson–Cook model. This method is able to calculate mean oblique cutting forces accurately through mechanics of machining. However, the tool-workpiece separation criterion in axial direction is relatively easy to decide, and the dynamics of moving tool or workpiece is simplified as a contact ratio. Khajehzadeh *et al.* [0] proposed theoretical model of tool temperature in ultrasonic vibration-assisted turning. The tool position is the superposition of linear feed movement and sinusoid vibration movement. The primary heat source in shear zone and secondary heat source at tool-chip interface are considered. However, the rubbing heat source on tool-workpiece interface is ignored. Gao and Sun [0] did analytical predictive modeling of surface roughness for two-dimensional vibration-assisted machining based on kinematics and tool geometry analysis, but the effects of forces and workpiece recovery are ignored. The residual stress and tool wear models have not been proposed in ultrasonic vibration-assisted milling.

1.2. Additive manufacturing

Laser-assisted metal additive manufacturing (LMAM) can produce geometrically complex parts with effective cost [0]. High-density laser power is employed to fully melt or fuse metal powders to build parts in a layer-by-layer manner. The LMAM processes can be broadly classified into two categories, namely powder bed metal additive manufacturing (PBMAM) and powder feed metal additive manufacturing (PFMAM). However, defects such as crack, porosity, undesired residual stress, and distortion are frequently observed in the parts produced by the LMAM process [0-0]. The large thermal gradient, due to the repeatedly rapid heat and solidification during the process, causes those common defects, as reported in the literature [0,0]. Different methods have been employed for temperature investigation. They are experimental technique, numerical modeling, and analytical modeling.

Experimental techniques include in-situ measurement and post-process measurement were employed for temperature investigation. The in-situ measurement techniques such as embedded thermocouple and infrared camera are widely used for temperature measurement in different manufacturing processes [0-0], which cannot achieve three-dimensional temperature measurement inside the build in LMAM processes [0,0]. The post-process measurement technique based on solidification microstructure measured under microscope required extensive experimental work for sample preparation [0]. The experimental techniques are limited by the restricted accessibility under elevated temperature levels.

Numerical models were developed based on the finite element method (FEM), which were computationally expensive. Roberts et al. developed a FEM model to predict the temperature distribution in multi-track, multi-layer scans using element birth and death technique [0]. Fu et al. developed another FEM model to predict the temperatures using solid material properties and powder material properties [0]. Similar models were employed for different LMAM processes with various materials [0]. Xia et al. developed a FEM model taking powder packing and powder size variation into consideration, which allows the prediction of build porosity [0]. FEM models were also employed for the prediction of residual stress and part distortion [0,0]. Although numerical models have made considerable progress on the prediction of the LMAM process, the expensive computational cost is still a major drawback.

Analytical models were developed based on the physics and heat transfer mechanism. Analytical models can calculate the three-dimensional temperature distribution in LMAM without resorting to FEM or any iteration-based simulations, and thus has promising short computational time. Van Elsen et al. summarized three moving heat source solutions, namely, point moving heat source solution, semi-ellipsoid moving heat source, and uniform moving heat source respectively [0]. The assumptions of the semi-infinite workpiece, and isotropic and homogeneous material were enforced in those models. The point moving heat source solutions was originally proposed by Carslaw and Jaeger [0]. Ning et al. further developed the point moving heat source to predict the temperature in multi-track scans considering the scanning strategy. The molten pool evolution was also calculated with temperature prediction [0]. Cline et al. developed another analytical model assuming a moving hear source with Gaussian distributed intensity profile [0]. This solution became the point moving heat source solution by reducing the beam radius to zero. Rosenthal developed a line moving heat source solution for welding process [0]. The semi-infinite medium was assumed in this solution. However, the developed analytical models cannot handle the heat transfer boundary condition, which resulted in unoptimized prediction accuracy, and thus prevent the accurate prediction for porosity, residual stress, and part distortion. The semi-analytical model was developed for temperature prediction, in which FEM was employed for imposing heat transfer boundary conditions [0]. The involvement of FEM unavoidably compromises the computational efficiency. In addition, Green's function was suggested by Carslaw and Jaeger for temperature prediction in a bounded medium. However, the complex mathematical calculations unavoidly reduce computational efficiency. To the knowledge of authors, a computationally efficient analytical model considering heat transfer boundary condition is not available yet.

2. Predictive modeling in subtractive manufacturing

The flow chart of the proposed predictive model is shown in Figure 1. Asterisk indicates step only in laser-assisted milling, while double asterisk indicates step only in ultrasonic vibration-assisted milling. At each rotation angle of milling tool, the cutting parameters and tool geometry parameters are recalculated as in equivalent orthogonal cutting. Then, the milling forces are calculated through flow stress based on mechanics of machining [0,0]. The forces also result in heat sources in machining temperature prediction [0]. And the residual stress is predicted based on mechanical stress from force and thermal stress from temperature field, followed by relaxation process [0-0]. The surface roughness is calculated considering the tool deformation under forces, tool path, and the elastic recovery of machined surface [0,0]. Lastly, the tool flank wear rate is decided by abrasive, adhesive, and diffusive wear based on the stress and temperature on tool-workpiece interface [0,0]. For laser-assisted milling, the laser preheating temperature field is predicted which affects the flow stress and overall temperature field [0]. In addition, the high temperature could trigger the microstructure evolution, which leads to grain growth and change of flow stress. For ultrasonic vibration-assisted milling, tool-workpiece separation criteria are examined first, and the prediction is only continued when there is contact between tool and workpiece at the moment [0].

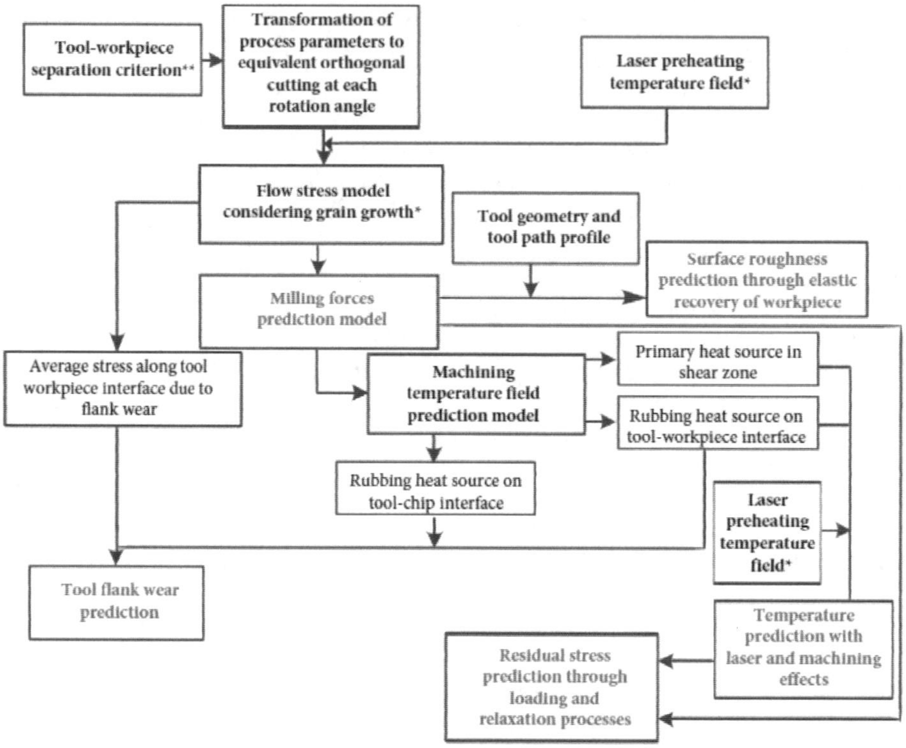

Figure 1. The flow chart of the predictive model in subtractive manufacturing.

2.1. Laser-assisted milling

The laser preheating temperature increase rate at top surface is described by

$$\Delta T_{\text{laser}}(x,0) = \frac{q(x) - h(T_{\text{laser}}(x,0) - T_0)}{\rho c_{\text{p}}} \quad (1)$$

where T_{laser} is the laser preheating temperature, ρ is the material density, c_{p} is specific heat, h is the heat transfer coefficient, T_0 is the environment temperature, $q(x)$ is the heat generation rate due to laser power described by the Gaussian equation as

$$q(x) = \frac{2Q}{\pi r^2}\exp(-\frac{2x^2}{r^2}) \quad (2)$$

where Q is the total input power of laser, r is the radius of laser spot. Within the workpiece, heat conduction is the dominant effect with the governing equation of

$$\Delta T_{\text{laser}}(x,z) = \alpha \nabla^2 T_{\text{laser}}(x,z) - V_{\text{f}} \frac{\partial T_{\text{laser}}(x,z)}{\partial x} \quad (3)$$

where α is thermal diffusivity and V_{f} is the moving speed of laser spot or the feed speed. The first term on the right describes the two-dimensional heat conduction, while the second term considers the effect of moving laser beam in cutting direction.

The recrystallization effect is considered through modified Johnson-Cook flow stress model as

$$\sigma = (A_{hp} + K_{hp} d^{-0.5} + B\bar{\varepsilon}^n)(1 + C\ln\frac{\dot{\bar{\varepsilon}}}{\dot{\bar{\varepsilon}}_0})\{1 - (\frac{T - T_0}{T_m - T_0})^m\} \quad (4)$$

where A_{hp}, K_{hp}, B, C, m and n are material constants, $\bar{\varepsilon}$ is the plastic strain, $\dot{\bar{\varepsilon}}$ is the plastic strain rate, $\dot{\bar{\varepsilon}}_0$ is the reference strain rate, and T_m is the melting temperature. The average grain size d is

$$d = d_{drex} X_{drex} + d_0 (1 - X_{drex}) \tag{5}$$

where d_0 is the initial average grain size, d_{drex} is the dynamically recrystallized average grain size and X_{drex} is the recrystallized volume fraction.

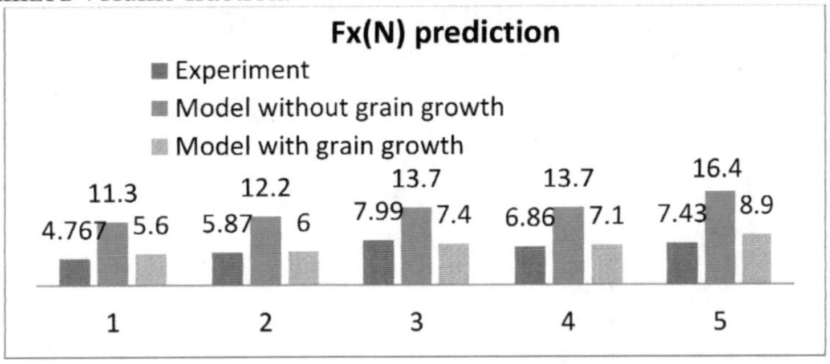

Figure 2. Comparison of Fx in laser-assisted milling [0].

As shown in Figure 2, with the consideration of microstructure evolution and grain growth, the error between measurements and predictions is less than 20% during laser-assisted milling of Inconel 718. As shown in **错误!未找到引用源。**, good agreement is found between measurements and predictions for residual stress after laser-assisted milling of Ti6Al4V. And the residual stress profile moves toward less compressive direction due to thermal softening effect.

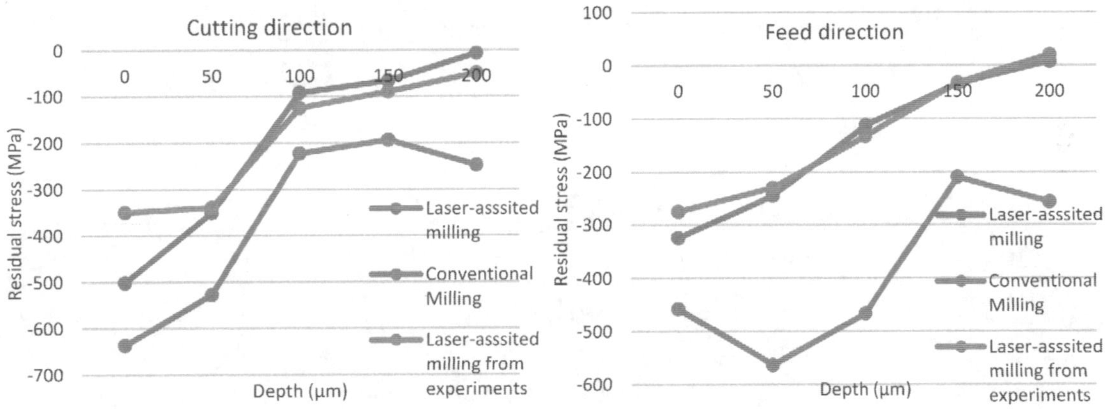

(a) (b)

Figure 3. Residual stress comparison for Ti6Al4V in (a) Cutting direction and (b) Feed direction [0].

2.2. *Ultrasonic vibration-assisted milling*

Kinematic analysis is conducted to check the tool-workpiece separation criteria. The velocity of tool tip is described by

$$V_x(t) = \omega A_x \cos(\omega t) - f - V_r \cos(\phi_r); \quad V_y(t) = \omega A_y \cos(\omega t) + V_r \sin(\phi_r) \tag{6}$$

where V_r is the cutting speed dependent on spindle rotation frequency and cutter size, ϕ_r is rotation angle, f is the feed rate, $\omega = 2\pi f_0$, f_0 is the ultrasonic vibration frequency, A_x ,and A_y are the vibration amplitude in each direction. There is no contact if the cutting direction component of the relative velocity between tool and workpiece is opposite to the tool rotation direction [0,0]. As shown in Figure 4(a),

$V_n = \sqrt{V_x^2 + V_y^2}$ is the resultant cutting velocity based on Equation (6), while $V_{ul} = z'(t)$ is the ultrasonic vibration velocity in axial direction, β is the helix angle. In Figure 4(b), the component along the transverse direction decides the intermittent effect at it is perpendicular to the uncut surface. When this component is negative, which is oppose to the tool rotation direction, there is no contact between tool and workpiece as described by

$$V_n \cos(\beta) - V_{ul} \sin(\beta) < 0 \tag{7}$$

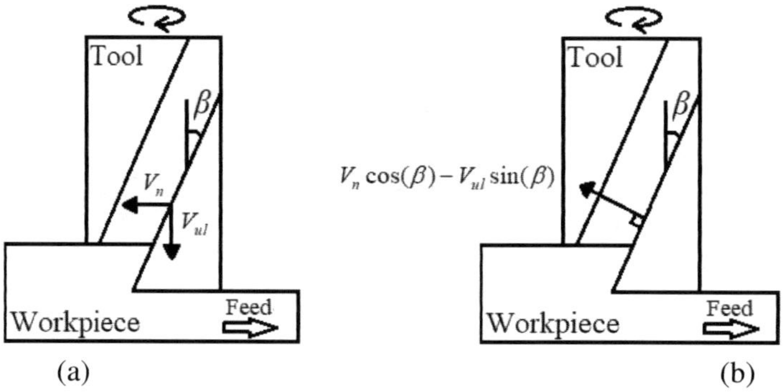

Figure 4. Velocity at the tool tip due to tool rotation and ultrasonic vibration: (a) representation of V_n and V_{ul} at the cutting tip (b) resolved component at the tip.

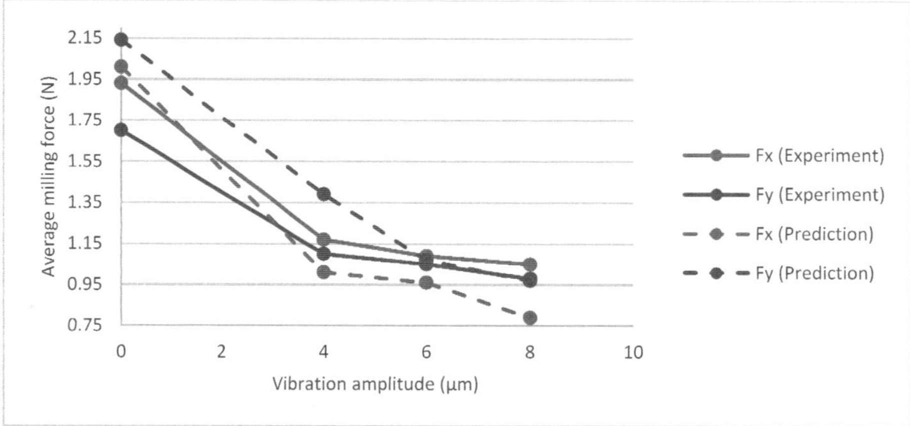

Figure 5. Comparison of average milling forces with different vibration amplitudes [0].

According to the average cutting forces from experiments and predictive model on Aluminum alloy 2A12 as shown in Figure 5, the average values decrease by 39% in feed direction and 35% in cutting direction when the ultrasonic vibration is applied. In addition, the measured forces keep dropping gradually when the amplitude increases, since the tool-workpiece separation time is longer under higher vibration amplitude. Overall, the proposed force prediction model is able to match the trend with average error of 13.6% in F_x and 13.8% in F_y.

3. Predictive modeling in additive manufacturing

3.1. Methodology

Ning et al. presented an analytical model for temperature prediction in metal additive manufacturing with consideration of heat transfer boundary [0, 0]. The heat transfer mechanism in PBMAM is illustrated as in Figure 6, in which the red arrow and green arrows denote the heat input from the laser, and heat loss from convection and radiation at the part boundary respectively. The build porosity, residual stress, and distortion can be predicted based on the temperature investigation, which are discussed as future works.

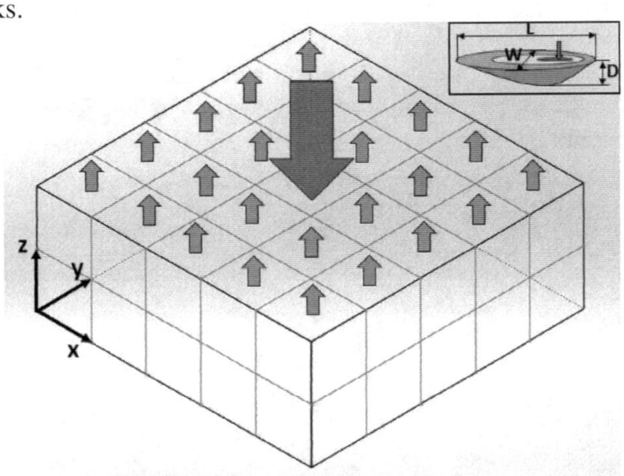

Figure 6. Heat transfer mechanism in powder bed metal additive manufacturing. The red arrow denotes heat input from the moving laser source. The green arrows denote heat loss from the convection and radiation at the part boundary. L, W, D denote molten pool length, width, and depth respectively [0].

The steady-state point moving heat source solution was originally proposed by Carslaw and Jaeger [0]. The assumptions of a semi-infinite medium, and isotropic and homogeneous material were enforced in this solution. It is expressed as the following

$$\theta_{laser}(x,y,z) = \frac{P\eta}{4\pi kR(T_m - T_0)} exp\left(\frac{-V(R+x)}{2\kappa}\right) \quad (8)$$

where P is laser powder, η is absorption, V is heat source moving velocity along x the direction, κ is thermal diffusivity ($\kappa = k/\rho c$, where k, ρ, c are thermal conductivity, density, and specific heat respectively), R is the distance from laser heat source ($R^2 = x^2 + y^2 + z^2$, x, y, z are the corresponding distances from the laser source respectively), θ is dimensionless temperature ($\theta = (T - T_0)/(T_m - T_0)$, T_m, T_0 are the material melting temperature and room temperature respectively).

The heat sink solution was derived with an equivalent power for heat loss from convection and radiation, and zero moving velocity. The convection and radiation can be calculated as

$$Q_{conv} = Ah(T - T_0) \qquad Q_{rad} = A\varepsilon\sigma(T^4 - T_0^4) \quad (9)$$

where Q_{conv} and Q_{rad} denote heat loss due to convection and radiation respectively, A is the area of heat sink, h is convection coefficient, ε is emissivity, σ is Stefan-Boltzmann constant, T is the temperature of the heat sink that can be estimated by the point moving heat source solution, T_0 is room temperature.

The equivalent power for heat loss at the part boundary can be expressed as

$$P_{equiv} = Ah(T - T_0) + A\varepsilon\sigma(T^4 - T_0^4) \quad (10)$$

The part boundary is mathematically discretized into many sections (heat sinks) considering the non-uniform temperature distribution at the part boundary, which causes non-uniform heat loss at the boundary.

$$\theta_{loss}(x,y,z) = \sum_{i=0}^{n} \frac{A_i}{4\pi\kappa R(T_m - T_0)} [h(T_i - T_0) + \varepsilon\sigma(T_i^4 - T_0^4)] \quad (11)$$

where n is the number of heat sinks, i is the index of each heat sink.

The final temperature solution is constructed from the superposition of heat source solution and heat sink solutions as the following

$$\theta_{laser}(x,y,z) = \frac{P\eta}{4\pi kR(T_m - T_0)} exp\left(\frac{-V(R+x)}{2\kappa}\right) - A\sum_{i=0}^{n} \frac{h(T_i - T_0) + \varepsilon\sigma(T_i^4 - T_0^4)}{4\pi\kappa R(T_m - T_0)} \quad (12)$$

In addition, the latent heat (L_f) was considered using heat integration method [0], in which the temperature were lowered because of phase transformation with continuous heat input.

$$\Delta T = L_f/c \quad (13)$$

3.2. Results and discussion

The temperature profiles were calculated by the presented model in PBMAM of Ti6Al4V. The thermal physics properties of Ti6Al4V are given in Table 1. It should be noted that the temperature-dependent material properties used only for the investigation of material properties variation. A 6 by 6 heat sink setting were determined based on the calibration with experimental measurement of the molten pool under test 1 process condition.

Table 1. Materials properties of Ti6Al4V [0,0].

Name	Symbol	Value	Unit
Density	ρ	4428	kg/m3
Thermal conductivity (powder at T_0)	k_p	6.6	W/(m·K)
Thermal conductivity (solid)	k_s	$-0.797 + 18.2 \times 10^{-3}T - 2 \times 10^{-6}T^2$ (T<1923K) 33.4 (T>1923K)	W/(m·K)
Specific heat (powder at T_0))	c_p	580	J/(kg·K)
Specific heat (solid)	c_s	$411.5 + 2 \times 10^{-1}T - 5 \times 10^{-7}T^2$ (T<1923K) *830* (T>1923K)	J/(kg·K)
Latent heat	H_f	365000	J/kg
Room temperature	T_0	20	°C
Solidus temperature	T_s	1605	°C
Liquidus temperature	T_l	1655	°C
Heat convection coefficient	h	24	W/(m²·K)
Emissivity	ε	0.9	1
Stefan-Boltzmann constant	σ	5.67 10-8	W/(m²·K⁴)
Absorption	η	0.77	1

Note: the temperature-sensitive material properties are only used for the investigation of material property variation.

The three-dimensional temperature profile was illustrated in Figure 7, in which the temperature profiles (solid lines) were compared to those calculated without considering heat loss (dashed lines). The temperature affected material property variations were shown in Figure 8, a) for thermal conductivity, b) for specific heat. The molten pool dimensions were determined by comparing the calculated temperature with the materials melting temperature. The molten pool dimensions were calculated under different process conditions and validated to the documented experimental values based on the solidification microstructure. As shown in Table 2, close agreements were observed upon validation to the experimental values. The computational time using a personal computer was recorded and presented.

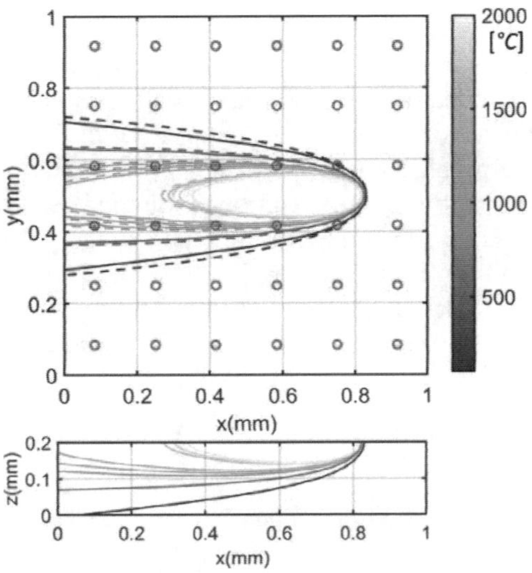

Figure 7. Calculated temperature profiles under test 1 process condition. Red circles denote the centers of heat sinks. For comparison, the temperature profiles (dashed lines) were calculated without the consideration of heat transfer boundary condition. The temperature profiles (solid lines) were calculated with consideration of heat loss from convection and radiation at the part boundary.

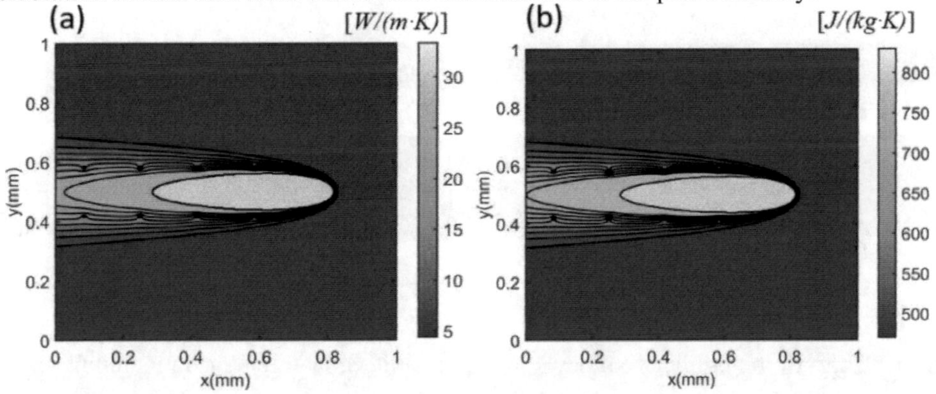

Figure 8. Calculated temperature-affected material properties (a) thermal conductivity (b) specific heat.

Table 2. Process conditions, experimental molten pool dimensions, calculated molten pool dimensions [0].

Test	Power (W)	Scanning Velocity (mm/s)	Molten pool width (μm)	Molten pool depth (μm)	Molten pool length* (μm)	Molten pool width* (μm)	Molten pool depth* (μm)	Computation time* (s)
1	100	500	118	62	530	120	63	93.73
2	100	750	98	52	537.5	100	50	88.52
3	100	1000	75	31	537.5	90	45	88.35
4	100	1200	72	36	537.5	80	40	88.74
5	150	500	146	122	752.5	145	113	90.33
6	150	750	136	88	787.5	125	83	91.10
7	150	1000	116	85	797.5	110	85	91.02
8	150	1200	108	59	800	100	60	89.76

Note: * denotes the calculated values using the presented model.

With the benefits of high computational efficiency and high prediction accuracy, the presented model can be used for temperature investigation for larger parts and process-parameters planning through inverse analysis [0-0]. Future works can be made to improve the presented model for powder feed metal additive manufacturing. The powder size distribution and powder packing should be considered because of their influence on material thermophysical properties.

Build porosity, residual stress and part distortion can be predicted based on the calculated temperatures as future works as illustrated in Figure 9. Some preliminary works on the predictions of build porosity and residual stress can be found in the references [0, 0]. The thermal stress was predicted by the thermoelasticity theory from the calculated temperatures. The residual stress was calculated by the elastoplastic contact algorithm and relaxation procedures. The build porosity can be predicted through regression analysis based on available experimental data.

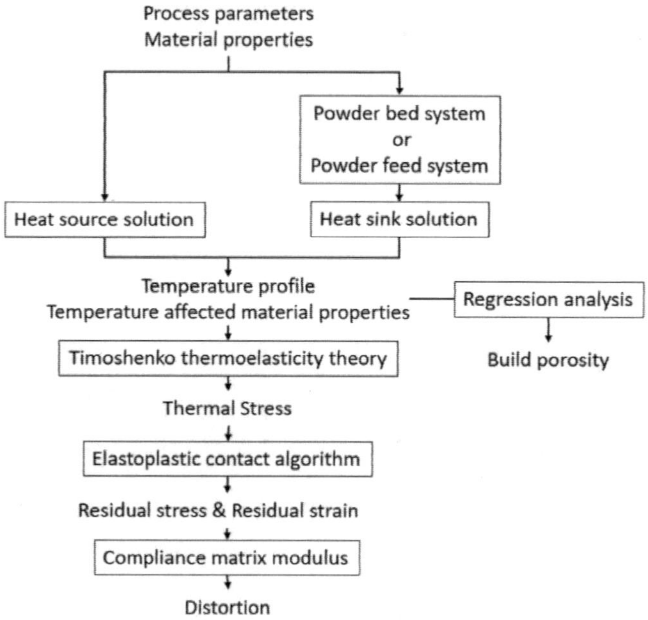

Figure 9. Flowchart of the predictive models in metal additive manufacturing including the presented model and future works.

4. Summary

In subtractive manufacturing, predictive modeling of laser-assisted and ultrasonic vibration-assisted milling is introduced. The effects of laser preheating and grain growth are considered in laser-assisted milling, while the intermittent tool-workpiece separation is considered in ultrasonic vibration-assisted milling. The preheating temperature field is predicted by considering the heat generation from laser and convection. The heat generation rate is described by Gaussian equation. The Hall-Petch equation is used to define a modified Johnson-Cook model parameter and account for grain growth effect. The tool-workpiece separation criterion is checked based on the instantaneous position and velocity of tool. Good agreements were found between measurements and predictions of force and residual stress in laser-assisted milling. And the force reduction in ultrasonic vibration-assisted milling was also accurately predicted.

In additive manufacturing, a closed-form solution was presented for temperature prediction in PBMAM configuration, without resorting to mesh of any iteration-based simulations. The temperature increase from the laser heat source and temperature decrease from the heat loss at the part boundary were calculated using point moving heat source solution and heat sink solutions respectively. The molten pool dimensions were calculated from the predicted temperature profile. Good agreements were observed upon validation to the documented experimental measurements. The high prediction accuracy allows the temperature prediction for large-scale parts and process-parameters planning through inverse analysis. The build porosity, residual stress, and part distortion can be further investigated based on the temperature investigation.

References:

[1] Bermingham, M.J., et al., Tool life and wear mechanisms in laser assisted milling Ti–6Al–4V. Wear, 2015. 322-323: p. 151-163.
[2] Halim, N.F.H.A., H. Ascroft, and S. Barnes, Analysis of Tool Wear, Cutting Force, Surface Roughness and Machining Temperature During Finishing Operation of Ultrasonic Assisted Milling (UAM) of Carbon Fibre Reinforced Plastic (CFRP). Procedia Engineering, 2017. 184: p. 185-191.
[3] Pan, Z., et al., Analytical model for force prediction in laser-assisted milling of IN718. The International Journal of Advanced Manufacturing Technology, 2017. 90(9): p. 2935-2942.
[4] Tian, Y., et al., Laser-Assisted Milling of Silicon Nitride Ceramics and Inconel 718. Journal of Manufacturing Science and Engineering, 2008. 130(3): p. 031013.
[5] Ulutan, D., B. Erdem Alaca, and I. Lazoglu, Analytical modelling of residual stresses in machining. Journal of Materials Processing Technology, 2007. 183(1): p. 77-87.
[6] Huang, Y., PREDICTIVE MODELING OF TOOL WEAR RATE WITH APPLICATION TO CBN HARD TURNING. 2002, Georgia Institute of Technology.
[7] Huang, Y. and S.Y. Liang, Modeling of CBN Tool Flank Wear Progression in Finish Hard Turning. Journal of Manufacturing Science and Engineering, 2004. 126(1): p. 98.
[8] Huang, Y., Y.K. Chou, and S.Y. Liang, CBN tool wear in hard turning: a survey on research progresses. The International Journal of Advanced Manufacturing Technology, 2007. 35(5-6): p. 443-453.
[9] Hao, Y. and Y. Liu, Analysis of milling surface roughness prediction for thin-walled parts with curved surface. The International Journal of Advanced Manufacturing Technology, 2017. 93(5-8): p. 2289-2297.
[10] Verma, G.C., P.M. Pandey, and U.S. Dixit, Modeling of static machining force in axial ultrasonic-vibration assisted milling considering acoustic softening. International Journal of Mechanical Sciences, 2018. 136: p. 1-16.
[11] Khajehzadeh, M. and M.R. Razfar, Theoretical modeling of tool mean temperature during ultrasonically assisted turning. Proceedings of the Institution of Mechanical Engineers, Part B: Journal of Engineering Manufacture, 2014. 230(4): p. 675-693.
[12] Gao, Y. and R.L. Sun. Modelling of Theoretical Surface Roughness for Two-Dimensional Vibration-Assisted Machining. in International Conference on Computer Information Systems and Industrial Applications. 2015. Atlantis Press.
[13] Frazier, W. E. (2014). Metal additive manufacturing: a review. Journal of Materials Engineering and Performance, 23(6), 1917-1928. https://doi.org/10.1007/s11665-014-0958-z
[14] Lewandowski, J. J., & Seifi, M. (2016). Metal additive manufacturing: a review of mechanical properties. Annual review of materials research, 46, 151-186. https://doi.org/10.1146/annurev-matsci-070115-032024
[15] Slotwinski, J. A., Garboczi, E. J., & Hebenstreit, K. M. (2014). Porosity measurements and analysis for metal additive manufacturing process control. Journal of research of the National Institute of Standards and Technology, 119, 494. https://doi.org/10.6028/jres.119.019
[16] Sames, W. J., List, F. A., Pannala, S., Dehoff, R. R., & Babu, S. S. (2016). The metallurgy and processing science of metal additive manufacturing. International Materials Reviews, 61(5), 315-360. https://doi.org/10.1080/09506608.2015.1116649
[17] Aboulkhair, N. T., Everitt, N. M., Ashcroft, I., & Tuck, C. (2014). Reducing porosity in AlSi10Mg parts processed by selective laser melting. Additive Manufacturing, 1, 77-86. https://doi.org/10.1016/j.addma.2014.08.001
[18] Paul, R., Anand, S., & Gerner, F. (2014). Effect of thermal deformation on part errors in metal powder based additive manufacturing processes. Journal of Manufacturing Science and Engineering, 136(3), 031009. https://doi.org/ 10.1115/1.4026524

[19] Ning, J., & Liang, S. Y. (2019). A comparative study of analytical thermal models to predict the orthogonal cutting temperature of AISI 1045 steel. The International Journal of Advanced Manufacturing Technology, 102(9-12), 3109-3119. https://doi.org/10.1007/s00170-019-03415-9
[20] Ning, J., & Liang, S. (2018). Prediction of temperature distribution in orthogonal machining based on the mechanics of the cutting process using a constitutive model. Journal of Manufacturing and Materials Processing, 2(2), 37. https://doi.org/10.3390/jmmp2020037
[21] Ning, J., & Liang, S. Y. (2019). Predictive Modeling of Machining Temperatures with Force–Temperature Correlation Using Cutting Mechanics and Constitutive Relation. Materials, 12(2), 284. https://doi.org/10.3390/ma12020284
[22] Ning, J., Nguyen, V., Huang, Y., Hartwig, K. T., & Liang, S. Y. (2019). Constitutive modeling of ultra-fine-grained titanium flow stress for machining temperature prediction. Bio-Design and Manufacturing, 1-8. https://doi.org/10.1007/s42242-019-00044-9
[23] Ning, J., & Liang, S. (2018). Evaluation of an analytical model in the prediction of machining temperature of AISI 1045 steel and AISI 4340 steel. Journal of Manufacturing and Materials Processing, 2(4), 74. https://doi.org/10.3390/jmmp2040074
[24] Heigel, J. C., Michaleris, P., & Reutzel, E. W. (2015). Thermo-mechanical model development and validation of directed energy deposition additive manufacturing of Ti–6Al–4V. Additive manufacturing, 5, 9-19. https://doi.org/10.1016/j.addma.2014.10.003
[25] Rodriguez, E., Mireles, J., Terrazas, C. A., Espalin, D., Perez, M. A., & Wicker, R. B. (2015). Approximation of absolute surface temperature measurements of powder bed fusion additive manufacturing technology using in situ infrared thermography. Additive Manufacturing, 5, 31-39. https://doi.org/10.1016/j.addma.2014.12.001
[26] Criales, L. E., Arısoy, Y. M., Lane, B., Moylan, S., Donmez, A., & Özel, T. (2017). Laser powder bed fusion of nickel alloy 625: experimental investigations of effects of process parameters on melt pool size and shape with spatter analysis. International Journal of Machine Tools and Manufacture, 121, 22-36. https://doi.org/10.1016/j.ijmachtools.2017.03.004
[27] Roberts, I. A., Wang, C. J., Esterlein, R., Stanford, M., & Mynors, D. J. (2009). A three-dimensional finite element analysis of the temperature field during laser melting of metal powders in additive layer manufacturing. International Journal of Machine Tools and Manufacture, 49(12-13), 916-923. https://doi.org/10.1016/j.ijmachtools.2009.07.004
[28] Fu, C. H., & Guo, Y. B. (2014). Three-dimensional temperature gradient mechanism in selective laser melting of Ti-6Al-4V. Journal of Manufacturing Science and Engineering, 136(6), 061004. https://doi.org/ 10.1115/1.4028539
[29] Romano, J., Ladani, L., & Sadowski, M. (2015). Thermal modeling of laser based additive manufacturing processes within common materials. Procedia Manufacturing, 1, 238-250. https://doi.org/10.1016/j.promfg.2015.09.012
[30] Xia, M., Gu, D., Yu, G., Dai, D., Chen, H., & Shi, Q. (2017). Porosity evolution and its thermodynamic mechanism of randomly packed powder-bed during selective laser melting of Inconel 718 alloy. International Journal of Machine Tools and Manufacture, 116, 96-106. https://doi.org/10.1016/j.ijmachtools.2017.01.005
[31] Mukherjee, T., Zhang, W., & DebRoy, T. (2017). An improved prediction of residual stresses and distortion in additive manufacturing. Computational Materials Science, 126, 360-372. https://doi.org/10.1016/j.commatsci.2016.10.003
[32] Afazov, S., Denmark, W. A., Toralles, B. L., Holloway, A., & Yaghi, A. (2017). Distortion prediction and compensation in selective laser melting. Additive Manufacturing, 17, 15-22. https://doi.org/10.1016/j.addma.2017.07.005
[33] Van Elsen, M., Baelmans, M., Mercelis, P., & Kruth, J. P. (2007). Solutions for modelling moving heat sources in a semi-infinite medium and applications to laser material processing. International Journal of heat and mass transfer, 50(23-24), 4872-4882. https://doi.org/10.1016/j.ijheatmasstransfer.2007.02.044
[34] Carslaw, H.; Jaeger, J. Conduction of Heat in Solids; Oxford Science Publication: Oxford, UK, 1990.

[35] Ning, J., Sievers, D. E., Garmestani, H., & Liang, S. Y. (2019). Analytical Modeling of In-Process Temperature in Powder Bed Additive Manufacturing Considering Laser Power Absorption, Latent Heat, Scanning Strategy, and Powder Packing. Materials, 12(5), 808. https://doi.org/10.3390/ma12050808

[36] Cline, H. E., & Anthony, T. (1977). Heat treating and melting material with a scanning laser or electron beam. Journal of Applied Physics, 48(9), 3895-3900. https://doi.org/10.1063/1.324261

[37] Rosenthal, D. The theory of moving sources of heat and its application of metal treatments. Transactions ASME 1946, 68, 849–866.

[38] Yang, Y., Knol, M. F., Van Keulen, F., & Ayas, C. (2018). A semi-analytical thermal modelling approach for selective laser melting. Additive Manufacturing, 21, 284-297. https://doi.org/10.1016/j.addma.2018.03.002

[39] Pan, Z., et al., Force modeling of Inconel 718 laser-assisted end milling under recrystallization effects. The International Journal of Advanced Manufacturing Technology, 2017. **92**(5): p. 2965-2974.

[40] Feng, Y., et al., Inverse analysis of the cutting force in laser-assisted milling on Inconel 718. The International Journal of Advanced Manufacturing Technology, 2018. **96**(1): p. 905-914.

[41] Feng, Y., Z. Pan, and S.Y. Liang, Temperature prediction in Inconel 718 milling with microstructure evolution. The International Journal of Advanced Manufacturing Technology, 2018. **95**(9): p. 4607-4621.

[42] Feng, Y., et al., Analytical and Numerical Predictions of Machining-Induced Residual Stress in Milling of Inconel 718 Considering Dynamic Recrystallization. 2018(51388): p. V004T03A023.

[43] Feng, Y., et al., Residual stress prediction in laser-assisted milling considering recrystallization effects. The International Journal of Advanced Manufacturing Technology, 2019. **102**(1): p. 393-402.

[44] Feng, Y., et al., Inverse Analysis of the Residual Stress in Laser–Assisted Milling. Preprints, 2019

[45] Feng, Y., et al., Surface roughness modeling in Laser-assisted End Milling of Inconel 718. Machining Science and Technology, 2019. **23**(4): p. 650-668.

[46] Feng, Y., et al., Inverse Analysis of Inconel 718 Laser-Assisted Milling to Achieve Machined Surface Roughness. International Journal of Precision Engineering and Manufacturing, 2018. **19**(11): p. 1611-1618.

[47] Feng, Y., et al., Flank tool wear prediction of laser-assisted milling. Journal of Manufacturing Processes, 2019. **43**: p. 292-299.

[48] Feng, Y., et al., Inverse analysis of the tool life in laser-assisted milling. The International Journal of Advanced Manufacturing Technology, 2019.

[49] Feng, Y., et al., Analytical prediction of temperature in laser-assisted milling with laser preheating and machining effects. The International Journal of Advanced Manufacturing Technology, 2019. **100**(9): p. 3185-3195.

[50] Feng, Y., et al., Inverse Analysis of the Residual Stress in Laser–Assisted Milling. Preprints, 2019

[51] Chen, W., et al., State-of-the-art review on vibration-assisted milling: principle, system design, and application. The International Journal of Advanced Manufacturing Technology, 2018. **97**(5-8): p. 2033-2049.

[52] Abdur-Rasheed, A., A Fundamental Study of Vibration Assisted Machining. Advanced Materials Research, 2011. **264-265**: p. 1702-1707.

[53] Ning, J., Mirkhooi, E., Dong, Y., Sievers, D. E., Garmestani, H., & Liang, S. Y. (2019). Analytical modeling of 3D temperature distribution in selective laser melting of Ti-6Al-4V considering part boundary conditions. Journal of Manufacturing Process, 44, 319-326. https://doi.org/10.1016/j.jmapro.2019.06.013

[54] Ning, J., Sievers, D. E., Garmestani, H., & Liang, S. Y. (2019). Analytical modeling of transient temperature in powder feed metal additive manufacturing during heating and cooling stages. Applied Physics A, 125(8), 496. https://doi.org/10.1007/s00339-019-2782-7

[55] de La Batut, B., Fergani, O., Brotan, V., Bambach, M., & El Mansouri, M. (2017). Analytical and numerical temperature prediction in direct metal deposition of Ti6Al4V. Journal of Manufacturing and Materials Processing, 1(1), 3. https://doi.org/10.3390/jmmp1010003

[56] Dilip, J. J. S., Zhang, S., Teng, C., Zeng, K., Robinson, C., Pal, D., & Stucker, B. (2017). Influence of processing parameters on the evolution of melt pool, porosity, and microstructures in Ti-6Al-4V alloy parts fabricated by selective laser melting. Progress in Additive Manufacturing, 2(3), 157-167. https://doi.org/10.1007/s40964-017-0030-2

[57] Ning, J., Nguyen, V., Huang, Y., Hartwig, K. T., & Liang, S. Y. (2018). Inverse determination of Johnson–Cook model constants of ultra-fine-grained titanium based on chip formation model and iterative gradient search. The international journal of advanced manufacturing technology, 99(5-8), 1131-1140. https://doi.org/10.1007/s00170-018-2508-6

[58] Ning, J., & Liang, S. Y. (2019). Inverse identification of Johnson-Cook material constants based on modified chip formation model and iterative gradient search using temperature and force measurements. The International Journal of Advanced Manufacturing Technology, 102(9-12), 2865-2876. https://doi.org/10.1007/s00170-019-03286-0

[59] Ning, J., & Liang, S. Y. (2018). Model-driven determination of Johnson-Cook material constants using temperature and force measurements. The International Journal of Advanced Manufacturing Technology, 97(1-4), 1053-1060. https://doi.org/10.1007/s00170-018-2022-x

[60] Fergani, O., Berto, F., Welo, T., & Liang, S. Y. (2017). Analytical modelling of residual stress in additive manufacturing. Fatigue & Fracture of Engineering Materials & Structures, 40(6), 971-978. https://doi.org/10.1111/ffe.12560

[61] Tapia, G., Elwany, A. H., & Sang, H. (2016). Prediction of porosity in metal-based additive manufacturing using spatial Gaussian process models. Additive Manufacturing, 12, 282-290. https://doi.org/10.1016/j.addma.2016.05.009

IOP Conf. Series: Materials Science and Engineering **842** (2020) 012025 doi:10.1088/1757-899X/842/1/012025

Research on FJSP Rescheduling Execution Cost Based on Modified Genetic Algorithm

Q C Zhou[1], Y K Huang[1], X L Xiong[2] and J Zhao[1]

[1]School of Mechanical Engineering, Tongji University, 1239 Siping Road, Shanghai, China
[2]Tongji Zhejiang college, No.168 Business Road, Zhejiang, China

Email: 576947181@qq.com

Abstract. In order to achieve dynamic production of smart plants, it is necessary to dynamically adjust schedule for disturbances in the manufacturing process. An executable optimized rescheduling plan needs developing, so the study of rescheduling cost is necessary. Based on the essential analysis of the rescheduling problem, a new flexible job shop rescheduling execution cost is proposed. The correctness of the execution cost is illustrated by the application in a modified genetic algorithm. The result shows that the rescheduling method with the execution cost as the fitness can generate a new plan with less deviation from the original schedule, which has a great significance for actual manufacturing.

1. Introduction

In the manufacturing process, scheduling is an indispensable task. Many scholars have studied scheduling under various working conditions and many results are valuable to the actual manufacturing. However, the uncertainties and disturbances in the manufacturing process make it difficult to achieve the desired results when the schedule is executed. How to deal with these inevitable uncertainties and disturbances is the subject of this paper.

This paper first solves the FJSP (Flexible Job-shop Scheduling Problem) through double coding genetic algorithm to get a pre-schedule. Then a rescheduling algorithm is built for device failure. By analyzing plant operation process, the essence of the rescheduling problem is discussed. And the composition of the rescheduling execution cost is analyzed as the objective function of the algorithm. Finally, an example is given to prove the feasibility of the algorithm.

2. Problem description

There were lots of researches about deterministic scheduling problems, providing important theoretical support for the manufacturing process, but the uncertainties in the manufacture is inevitable, such as uncertainty in working hours, equipment failure, emergency order and order cancellation, etc. Therefore, some scholars are studying how to deal with these uncertain factors. The main research directions are: 1. Research on rescheduling method after interference occurs; 2. How to generate pre-scheduling scheme with certain robustness; 3. Research on rescheduling strategy. Representative studies include: R. J. Abumaizar proposed a rescheduling method for job shop problems, using makespan, deviation of starting time and sequence as a measure of rescheduling plans [1]; Sanjay V. et al. proposed a predictive scheduling method, by properly inserting idle time between operations, a pre-schedule that can absorb interrupts can be formed without significantly affecting the work performance [2].

In recent years, research on production scheduling has focused on: improvement of algorithms, such as improving evolutionary algorithms through learning methods [3]; proposing new algorithms applied to scheduling, such as hybrid artificial bee colony algorithm [4], flower pollination algorithm, etc. [5]; combining scheduling with production practice, such as applying real-time manufacturing data in scheduling [6],[7].

The FJSP can be stated as follows. It is given a set $J = \{J_1, J_2 ... J_n\}$ of independent jobs. A job has a sequence of operations $\{O_{i1}, O_{i2} ... O_{in_i}\}$ to be performed one after the other according to the given sequence. It is given a set of machines $U = \{M_1, M_2 ... M_m\}$. Each operation O_{ij} can be executed on any among a subset of compatible machines $U_{ij} \subseteq U$. The processing time of each operation is machine-dependent. T_{ij}^{k} stands for the processing time of operation $O_{ij}$when executed on machine M_k. Suppose S^{ori} is the pre-schedule being executed. A new schedule after a disturbance is S^{re}.

The above problems can be well solved theoretically by various algorithms, but in many cases, the new schedules obtained by these algorithms have high execution cost or even are not executable. In order to get executable optimized results, the essence of rescheduling is discussed, providing a reference for the better application of rescheduling theory.

3. Rescheduling execution cost

The above problem description has simplified the problem, leading to a big difference between reality and theory. In order to reduce the deviation, it is necessary to mine the essence of the rescheduling problem.

In the manufacturing process, the production enterprise first determines the batches according to the production demand, and then expands the batches to the self-made parts according to the BOM to form the schedule of each period. And the master production schedule is generated after the capacity check.

The master production schedule is a plan to decide the quantity of each specific final product to be manufactured in each specific period, which is usually months or weeks. And the detailed production schedule is always specific to each hour. Detailed production schedule is formed based on the master production schedule, the manufacturing process, the capacity benchmark data and the working calendar.

The production scheduling problem is abstracted from the detailed production plan. The essence is to arrange working center to complete certain production tasks in certain periods. The optimization of scheduling is essentially the process of improving the production capacity of the working center.

When the production process is disturbed, the production capacity of the working center is reduced, and there are many ways to cope with insufficient capacity, including: 1. delaying the unfinished work to next period; 2. the working center working overtime; 3. requesting help from other departments; 4. change the self-made parts to the outsourcing. According to the above description, the rescheduling method can be interpreted as utilizing the extending working hours to cope with the disturbance. The rescheduling optimization process is essentially to maintain the capacity of the work center with minimal overtime when the disturbance occurs. Therefore, the overtime and any process reducing production capacity are considered as rescheduling costs.

Some scholars have already studied the rescheduling cost: Georgios et al. [8] believed that only using time as the objective function of rescheduling is a short-sighted method, which is easy to produce pseudo-optimal results that are not available for actual production, then the rescheduling cost consisting of starting time deviation, equipment change, and process sequence change is proposed; some other scholars also proposed ideas on the cost of rescheduling [9]. But they only enumerated the composition of the rescheduling cost, lacking the basis for their selection. What's more, they did not consider overtime as rescheduling costs, but differentiated overtime from rescheduling costs as a part of the rescheduling objective function, which is not reasonable enough from the perspective of this work.

Based on the above analysis, the rescheduling cost should include:

Completion Time Delay for Each Machine. The delay in the completion time reflects the overtime. Essentially, the rescheduling optimization process hopes that the overtime hours are as small as possible, and the overtime hours are negative in terms of manpower, energy consumption and economy. Overtime can be defined as following:

$$A_1 = \sum_{i \in U} (L_i^{re} - L_i^{ori}) \tag{1}$$

U is the collection of machines;
L_i^{ori} is completion time of machine i in the pre-scheduling plan;
L_i^{re} is completion time of machine i in the rescheduling plan;

Total Starting-time Deviation. The total starting-time deviation of all batches from pre-schedule was first introduced by S. David Wu as a schedule stability measure [10]. The biggest impact of the starting-time deviation is the material accumulation in each work unit. In the ideal case of JIT, the materials will arrive at the units just when they are needed, but under the practical circumstances, the work units will often storage the materials required for subsequent processes. If the deviation of the starting time is large, the accumulation of materials will result in insufficient space for the working process, which will affect the manufacturing process. Sometimes some materials need shipping back to the warehouse, which increase the logistics cost. If the preparation amount in the units is large, the weight of this part in execution cost needs to be increased during the rescheduling process.

$$A_2 = \sum_{i \in J} \sum_{j \in O_i} |ST_{ij}^{re} - ST_{ij}^{ori}| \tag{2}$$

J is a collection of jobs to be processed;
O_i is a set of operations for the job i;
ST_{ij}^{ori} is the starting-time of the operation j of the job i in the pre-scheduling plan;
ST_{ij}^{re} is the starting-time of the operation j of the job i in the rescheduling plan;

Logistics Cost Due to Machine Reallocation. Machine reallocation for an operation changes the material requirements of the work unit. Materials prepared in advance at the work units do not match actual demand. The possible consequences include the transfer of materials between the units, the materials need to be replenished from the warehouse, and even the production may stagnate, which indirectly reduce production capacity.

$$A_3 = \sum_{i \in J} \sum_{j \in O_i} D(eq_{ij}^{ori}, eq_{ij}^{re}) \tag{3}$$

D is a matrix of distances between units;
eq_{ij}^{ori} is the device selected for j process of job i in the pre-scheduling plan;
eq_{ij}^{re} is the device selected for j process of job i in the rescheduling plan;

Processing Time Change Due to Machine Reallocation. The increase in processing time obviously reduces production capacity.

IOP Conf. Series: Materials Science and Engineering **842** (2020) 012025 doi:10.1088/1757-899X/842/1/012025

$$A_4 = \sum_{i\in J}\sum_{j\in O_i}(PT_{ij}^{ori} - PT_{ij}^{re}) \tag{4}$$

PT_{ij}^{ori} is the processing time of the j process of the job i in the pre-scheduling plan;
PT_{ij}^{re} is the processing time of the j process of the job i in the rescheduling plan;

The Rescheduling Cost can be expressed as:

$$A = \omega_1 A_1 + \omega_2 A_2 + \omega_3 A_3 + \omega_4 A_4 \tag{5}$$

$\omega_1, \omega_2, \omega_3, \omega_4$ are the weights of rescheduling execution cost components.

The above-mentioned rescheduling cost components are difficult to normalize, so the weight of each component must be formulated according to the actual situation of the production enterprise.

4. Algorithm construction

For the problem of flexible job shop, scholars have developed various effective algorithms, such as: Tabu search algorithm, genetic algorithm, simulated annealing algorithm, etc.

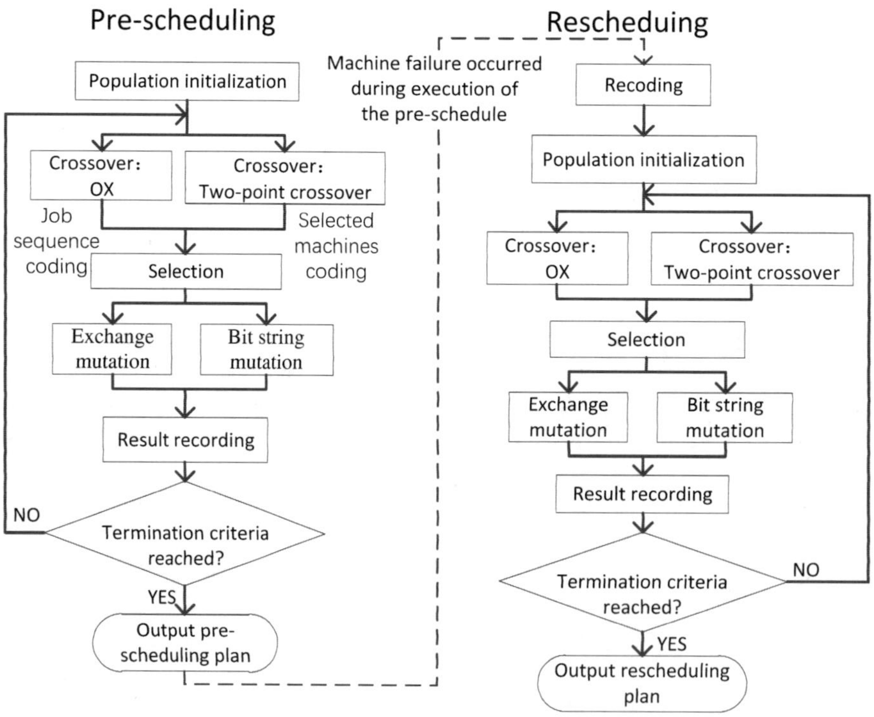

Figure 1. Algorithm block diagram.

Genetic Algorithm (GA) is a metaheuristic inspired by the process of natural selection that belongs to the larger class of evolutionary algorithms (EA). Genetic Algorithm has been used extensively "as a powerful tool to solve various optimization problems such as integer nonlinear problems (INLP)" [11]. Therefore, an improved genetic algorithm is used as an example of rescheduling cost application.

Scheduling is derived by modified genetic algorithm as shown in the figure 1, and a double coding method is used here, one chromosome represents the job sequence, and the other chromosome represents the selected device. The fitness of the algorithm is makespan, which is replaced by rescheduling cost in the rescheduling process. When machine M_i fails at time t, it is estimated that it takes T hours to repair. Rescheduling is triggered, and the chromosomes need recoding. The part that has been executed and is being executed needs to be removed in the chromosomes. The initial operation time MT_i of machines should be modified.

5. Case analysis

Apply the algorithm to an 6×5 FJSP example, which means 6 jobs need to be machined, and each job has 5 operations. Equipment available for each operation and processing time are shown in the Table 1 ([Available machines]/ [Processing time]). Assuming the units are linearly arranged by machine number, the machine number difference is used as the distance metric of the two units.

Table 1. Available machines and the processing time.

	O_1	O_2	O_3	O_4	O_5
J_1	[1,2]/ [3,4]	[2]/ [8]	[3,4]/ [9,7]	[2,4]/ [5,4]	[2,5]/ [3,3]
J_2	[1]/ [6]	[2,4]/ [8,6]	[3]/ [4]	[3,4]/ [2,6]	[5]/ [3]
J_3	[3]/ [4]	[1,3]/ [5,7]	[2]/ [7]	[1,2]/ [5,5]	[4,5]/ [6,6]
J_4	[5]/ [7]	[2]/ [3]	[2,4]/ [4,6]	[5]/ [3]	[2,5]/ [3,7]
J_5	[1,2]/ [6,4]	[3]/ [7]	[4,5]/ [7,9]	[1]/ [8]	[3,4]/ [5,5]
J_6	[2,3]/ [3,7]	[4]/ [7]	[3]/ [6]	[3,4]/ [6,7]	[2,4]/ [4,4]

First, pre-schedule the operations that need to be processed. The convergence curve can be seen in figure 2, and the optimal individual makespan is stable at 39h after 50 generations. Pre-schedule is as shown in the (b) Gantt chart. Assume that the fault occurred on machine 2 at the time of 15h, and the estimated recovery time is 8 hours. Rescheduling by two methods, one is rescheduling with makespan as fitness, and the other is rescheduling with execution cost as fitness, the result is shown in figure (c) (d) (e) (f) of figure 2.

Comparing to right-shift rescheduling, both above methods have better results, and the makespan of both rescheduling methods are the same. A labeling method is used to show the difference between the rescheduling plan and the pre-scheduling plan. As shown in 2-2 3-2 of figure 2, change of starting-time, operation machine and processing time in 2-2 is significantly more than 3-2. That means rescheduling with execution cost as fitness forms a new schedule with minimal changes to the original plan. It maintains production capacity in an executable and optimized way when the disturbances occur.

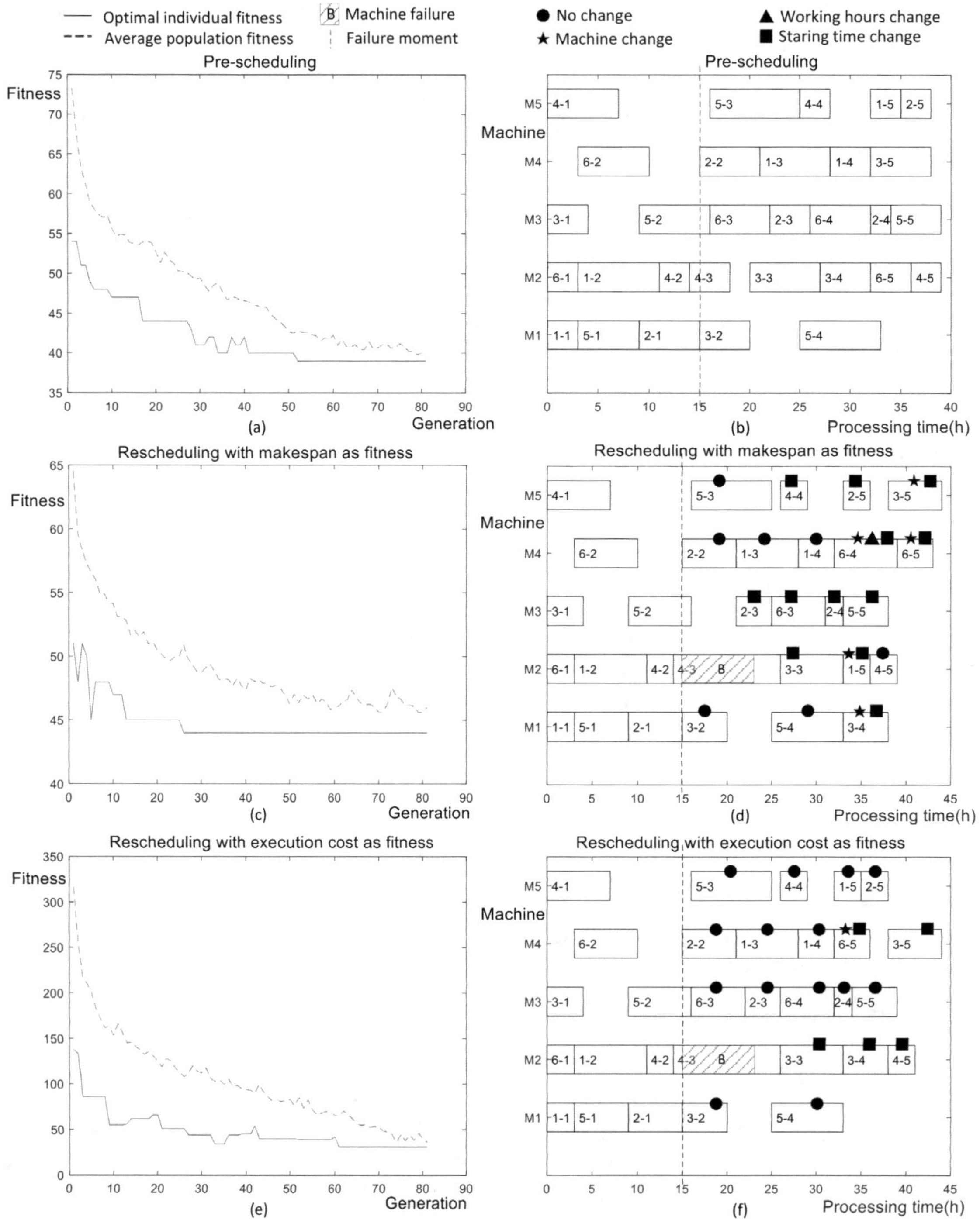

Figure 2. Convergence curves and gantt charts.

6. Conclusion

The rescheduling of FJSP is studied in this work, and a modified genetic algorithm is used to solve the problem. The rescheduling execution cost is mainly discussed, and the viewpoint is presented: the rescheduling optimization process is essentially to maintain the capacity of the work center with minimal overtime when the disturbance occurs. According to this viewpoint, the components of the rescheduling execution cost is analyzed, which consists of 1. Completion time delay for each machine, 2. Total starting-time deviation, 3. Logistics cost due to machine reallocation, 4. Processing time change due to machine reallocation. The feasibility of the algorithm and the rescheduling execution cost are proved by an example.

The ideal of production management is to realize real-time optimal operation of dynamic production. This work provides a viable method for the ideal realization.

7. References

[1] Abumaizar R J, Svestka J A. Rescheduling job shops under random disruptions[J]. *International journal of production research*, 1997, **35**(7): 2065-82.

[2] Mehta S V, Uzsoy R M. Predictable scheduling of a job shop subject to breakdowns[J]. *IEEE Transactions on Robotics and Automation*, 1998, **14**(3): 365-378.

[3] Lin L, Gen M. Hybrid evolutionary optimisation with learning for production scheduling: state-of-the-art survey on algorithms and applications[J]. *International Journal of Production Research*, 2018, **56**(1-2): 193-223.

[4] Li X, Peng Z, Du B, et al. Hybrid artificial bee colony algorithm with a rescheduling strategy for solving flexible job shop scheduling problems[J]. *Computers & Industrial Engineering*, 2017, **113**: 10-26.

[5] LIU Erhui, YAO Xifan, Tao Tao, JIN Hong. Improved Flower Pollinaton Algorithm for Job Shop Scheduling Problems Integrated with AGVs[J]. *Computer Integrated Manufacturing Systems*:1-38[2018-12-02].

[6] Zhang Y, Huang G Q, Sun S, et al. Multi-agent based real-time production scheduling method for radio frequency identification enabled ubiquitous shopfloor environment[J]. *Computers & Industrial Engineering*, 2014, **76**: 89-97.

[7] ZHANG Zequn, TANG Dunbing JING Yongqiao, ZHANG Haitao. Self-organizing Production Technology for Discrete Workshop Scheduling Driven by Internet of Things [J]. *Journal of mechanical Engineering*, 2018,**54**(16):34-44.

[8] Kopanos G M, Capón-García E, Espuna A, et al. Costs for rescheduling actions: a critical issue for reducing the gap between scheduling theory and practice[J]. *Industrial & Engineering Chemistry Research*, 2008, **47**(22): 8785-95.

[9] Liu Mingzhou, Wu Wenrui, Zhang Mingxin, et al. Reach on Rescheduling Excution Costs in Job Shop under Uncertain Conditions[J]. *China Mechanical Engineering* **24.6** (2013): 765-769.

[10] Wu S D, Storer R H, Pei-Chann C. One-machine rescheduling heuristics with efficiency and stability as criteria[J]. *Computers & Operations Research*, 1993, **20**(1): 1-14.

[11] Sadeghi J, Niaki S T A, Malekian M R, et al. Optimising multi-item economic production quantity model with trapezoidal fuzzy demand and backordering: two tuned meta-heuristics[J]. *European Journal of Industrial Engineering*, 2016, **10**(2): 170-195.

IOP Conf. Series: Materials Science and Engineering **842** (2020) 012026 doi:10.1088/1757-899X/842/1/012026

Fog-based Logistic Application Modeling in an Industry 4.0 Framework

Imen BOUZARKOUNA[1,*], M'hammed SAHNOUN[1,*], David BAUDRY[1], Mohamed Amine BENATIA[1] , Christian GOUT[2]

[1] 685 Avenue de l'Université, 76800 Saint-Étienne-du-Rouvray, France
[2]80 rue Edmund Halley Rouen Madrillet Innovation, 76800 Saint-Étienne-du-Rouvray, France

[*] E-mail: ibouzarkouna@cesi.fr, msahnoun@cesi.fr

Abstract. The industrial growth introduced the Industry 4.0 revolution, which uses several IoT-based technologies to manage the huge amount of data generated by its connected systems. This revolution aims to exchange data between the different parts and improve the performance of the manufacturing process. Facing several limitations, this paradigm needs the Fog Computing to ameliorate the Cloud services and the distribution of real-time decision-making and the secure and interoperable data analysis. The internal logistics presents the need to transmit and treat the data in real-time to take a more efficient decisions in a shorter delay. The Fog-based architecture is composed of terminal devices, Edge, Fog and Gateway devices. A model is presented to optimize the Fog integration (The cost of the links and nodes). Experimental results of an exact method optimization using Fico Xpress show a reduction of the Fog Computing integration cost in an internal logistic center.

1. Introduction

The increasing complexity of production systems and quantity of data is due to several factors such as sensor networks integration, software applications, variability of markets, etc. As a result, the emergence of a fourth revolution does not only strengthen the competitive position of developed countries, but also generates solutions to global and specific challenges such as energy, resource efficiency, complexity of systems and demographic changes management [1]. The Industry 4.0 emergence is becoming a necessity to satisfy the demand of process automation, integration of artificial intelligence, flexibility and high products customization.

The Industry 4.0 is based on several components like IoT (Internet of Things), the CPS (Cyber Physical Systems), Cloud Computing and Big Data [2]. These components lead the Industry 4.0 to another level of intelligence, automation and interoperability.

The Industry 4.0 requires computing data and making decision in real-time, however the Cloud Computing features are not adequate to transmit the information quickly [4]. Several new solutions able to increase the information pertinence and reduce time delay and computing consumption are developed to overcome the Cloud Computing lacks [5]. ©Cisco Systems has presented the "Fog Computing" paradigm as the most adequate solution for this issue. The Fog Computing is an Edge Computing solution, defined as the highly virtualized platform that extends the Cloud services and processes information directly at edge of the network [4], to accomplish processing, computing, storage, and routing services between traditional Cloud terminals and servers. The adoption of the Fog in the industry

4.0 is limited by numerous challenges (Cost, power consumption, interoperability, security, etc.,) that needs to be overcome [6].
One of the most influencing challenges is the cost of the Fog solution. Several industrial applications can be ameliorated by adopting a Fog-Computing solution to reduce delay and cost. One of these application is the internal logistic. This paper presents the importance of integrating the Fog Computing in the internal logistic application especially in the Industry 4.0 framework and defines its architecture composed of different layers. It also presents a model that decides whether to deploy or not the different architecture nodes, which optimize the cost of integrating the Fog Computing solution.
The paper is structured as follows: The next section (Section 2) presents an overview of the internal logistic in the Industry 4.0 and the existing works. The Section 3 presents the Fog-based architecture and the cost optimization model of an internal logistic Fog-based application. This section gives also the results of the model optimization. The paper ends with a conclusion 5 on the future directions of research for enhancing the model and using its optimal solution on a real use case that deploys the Fog in an internal logistic application.

2. Internal logistic in Industry 4.0

While the Cloud offers virtually high computational resources making data processing extremely efficient, the resulting data movement could have an impact on the application performance due to possibly significant latencies of the transmission. To improve this type of applications, proper information logistics become fundamental for delivering information at the right time, the right place, and with the right quality [7]. The Industry 4.0 afford smart services and smart products to final users The logistics, as an important part of the industrial process, needs to adapt to the succession of revolutions and the changing industrial vision. The logistic process inside the workshop involves numerous types of connected systems such as AGVs (automated guided vehicles, which follow path lines to transport materials and products around the industrial workshop), machines, manual posts, etc. The ordinary manners and technologies that support industrial logistics are no more enough for the Industry 4.0 requirements, therefore integrating the Fog features is beneficial [8]. Considering the importance of such application to organize the transfer, the industry 4.0 requires that logistic services need to become more and more smart, efficient and real-time reacting to respond to the high artificial intelligence level. Adding the Fog Computing characteristics will help taking quickly the most adequate decision to ameliorate the logistic operations. Connecting the Fog with the Cloud will deliver more analysis that manage future logistic tasks.
To integrate the Fog Computing, many aspects need to be studied especially modeling the application architecture and optimizing the cost of such solution and considering the latency, distribution and capacity limitations. Some works interested in this research domain like [9], which presented the integration of the Cloud Manufacturing in the production logistic. Authors in [10] presented a Fog-based solution in the logistic centers.

3. Fog-based architecture modelling

Among numerous Industry 4.0 applications, we are interested in logistic application. This paper presents an example of deploying the Fog Computing architecture in an Industry 4.0 framework, which is illustrated in Figure 1. This architecture is composed of four layers communicating between each other's to ensure a distributed real-time data transmission and computing.

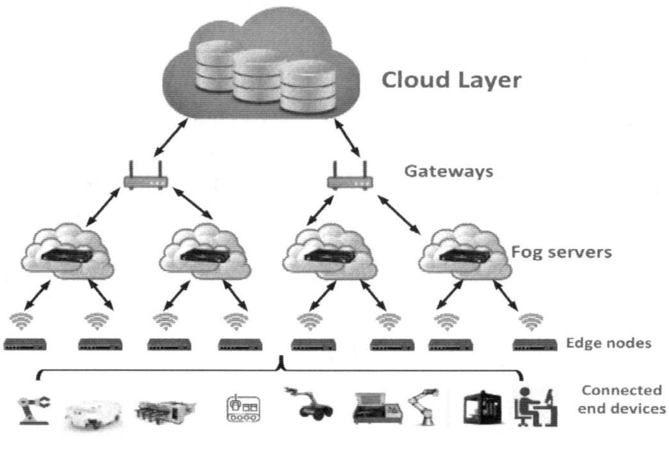

Figure 1. A Fog-based industrial architecture.

3.1. Model and constraints

Our work is based on the model proposed by [10] used in logistic centers in an Industry 4.0 context. An additional constraint "Networking capacity" is added to restrict a maximal bandwidth that transmission cannot exceed. This model aims to reduce the cost of links between Edge, Fog and Gateway and the cost of installing each node of them. The model will be solved using the exact optimization approach to activate the minimal number of links.
Minimize the objective function Z (1):

$$Z = c_f * \sum_{i\in C\cup\Omega_G\cup\Omega_F, j\in\Omega_G\cup\Omega_F\cup\Omega_E} x_{ij}\, d_{ij} + c_G * \sum_{c\in\Omega_G} G_c + c_F * \sum_{m\in\Omega_F} F_m + c_E * \sum_{n\in\Omega_E} E_n \quad (1)$$

With c_f is the cost of one unit of optical fiber for high-speed network, x_{ij} is the binary decision variable that decide if there is a link between nodes i and j, G_c, F_m *and* E_n are the binary decision variable deciding if the node is active or not. c_G, c_F and c_E are the cost of installing Gateway, Fog and Edge nodes respectively. d_{ij} is the distance between two nodes i and j. S is the AGVs set, C is the Cloud servers set and Ω_G is a c Gateway set, Ω_F is a m Fog set and Ω_E is a n Edge set.
Considering that the initial positions and the distances are fixed, activating nodes and establishing links will be according to several constraints. The first constraint is the activating requirements. Constraint (2), (3) and (4) is forcing that each AGV needs to be connected to one Edge and each edge must be activated if it is related to at least one AGV.

$$\sum_{n\in\Omega_E} x_{nk} = 1 \ ,\forall\, k \in S \quad (2)$$
$$E_n \le \sum_{k\in S} x_{nk} \ ,\forall\, n \in \Omega_E \quad (3)$$
$$x_{nk} \le E_n \ ,\forall\, n \in \Omega_E \ ,\forall\, k \in S \quad (4)$$

The same constraints are defined for the Edge-Fog and Fog-Gateway links. These constraints guarantee that each higher node will be connected to one and only lower node and activated when it is related at least to one lower node. The link between Cloud and Gateway depends on the same constraint.
The second constraint type is the capacity demand. Each higher node cannot accept more than it maximal computing capacity (5) and (6) (respectively for the Edge H_n^E, Fog H_m^F and Gateway H_c^G). And the capacity demand for an $\omega(E_n)$ is depending of the total of its connected lower nodes, but it can not exceed its maximum capacity E_{max} (respectively for the $\omega F(F_m)$ and F_{max}(7) and the $\omega G(G_c)$ and G_{max}(9). The r_{AGV} is the demand of each AGV. The same constraints (8) and (10) are established for Fog and Gateway links.

$$\omega(E_n) = \sum_{k\in S} r_{AGV} . x_{nk} \quad ,\forall n \in \Omega_E \tag{5}$$

$$\omega(E_n) \leq E_n . H_n^E \quad ,\forall n \in \Omega_E \tag{6}$$

$$\omega F(F_m) = \sum_{n\in\Omega_E} E_{max} . x_{mn} \quad ,\forall m \in \Omega_F \tag{7}$$

$$\omega F(F_m) \leq F_m . H_m^F \quad ,\forall m \in \Omega_F \tag{8}$$

$$\omega G(G_c) = \sum_{m\in\Omega_F} F_{max} . x_{cm} \quad ,\forall c \in \Omega_G \tag{9}$$

$$\omega G(G_c) \leq G_c . H_c^G \quad ,\forall c \in \Omega_G \tag{10}$$

Constraint (11) defines delay of each transmission needs to respect the maximal latency (for Edge γ_k^n, Fog γ_n^m and Gateway γ_m^c) that could be reached between the two levels.

$$L. E_n \leq \gamma_k^n . \left(\sum_{k\in S} x_{nk} . \mu_{kn} \right) ,\forall n \in \Omega_E \tag{11}$$

With L: data length, μ_{kn}: the data rate from an AGV to an Edge node. The same procedure is followed for Edge-Fog and Fog-Gateway links.

As the terminals are wireless connected and the AGVs are mobile, they need to be present in an Edge coverage zone to be attached. The constraint (12) defines the area where the robot ca be connected to the edge nod, with R_n is the Edge diameter.

$$\mathbf{x_{nk}} . \mathbf{d_{nk}} \leq \frac{\mathbf{R_n}}{2} \quad ,\forall \mathbf{n} \in \Omega_E, \forall \mathbf{k} \in \mathbf{S} \tag{12}$$

The constraint (13) requires that each Edge node cannot connect devices more than its link capacity N_E(Respectively N_F for Fog and N_G for Gateway), which distributes the terminal device on the active nodes.

$$\sum_{k\in s} x_{nt} \leq N_E \quad ,\forall \mathbf{n} \in \Omega_E \tag{13}$$

3.2. Optimization

The architecture is simplified and illustrated in Figure 2 to present the potential position of the Edge, Fog, Gateway and Cloud nodes. We supposed in the example that all the terminal nodes are AGVs, which are wirelessly connected with the Edge nodes. The AGVs insure the interne logistic in an Industry 4.0 workshop. The rest of the links are wired (using optical fiber). This scenario needs to be optimized to reduce the cost of this Fog-based architecture. The initial architecture is composed of 4 AGVs, 2 machines, 1 3D printer, 12 Edge, 7 Fog, 2 Gateway and 1 Cloud. The position of each node is fixed and the distance between each node is calculated depending on their link type (wireless/fiber). The model is solved using the exact method approach and the Linear Programming using simplex method. The model is optimized using the solver of Fico "Xpress IVE" to minimize the initial cost of the Fog-based solution. This solver is using Mosel language. The Figure 3 shows the results of the model optimization. As an input data, two distance matrix are prepared (a Manhattan distance for the wired links and a Euclidean distance for the wireless links) and all potential position of the different nodes and terminal devices according to their position on the plan. The Z (cost) is notably reduced from 3754.85 to 385.314.

4. Conclusion

The integration of the Fog Computing in the Industry 4.0 applications is the result of the increase use of the IoT that enables the communication between each other machines and humans in real time and the use of what is known as advanced digitalization. The internal logistic requires faster decision making and data transmission which is possible due to the Fog Computing capabilities. This paper presented the modeling and the optimization of a Fog-based solution for a logistic framework using AGVs and Fog servers, while considering several constraints (Links, latency, demand capacity and coverage). The results of the exact method solver shows an optimization in term of the Fog-based solution cost and reducing the number of active nodes. Future works are interested in ameliorating the current model and optimize the results of a larger scale Industry 4.0 example. The optimized model results will be also adopted in a real use case (Using the Cesi's prototype of a connected manufacturing system). This Fog servers will execute models to analyze data locally and provide more efficient information to the terminal devices and transmit data to the Cloud servers for more analysis and storage. Other challenges should be studied like the data security and the interoperability between the different existing operating systems.

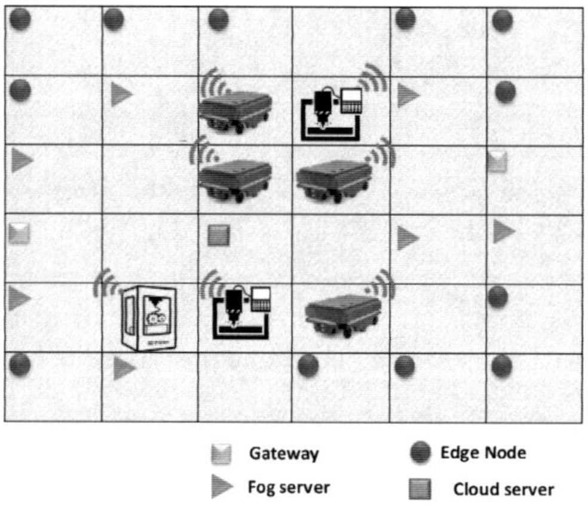

Figure 2. Potential nodes positions.

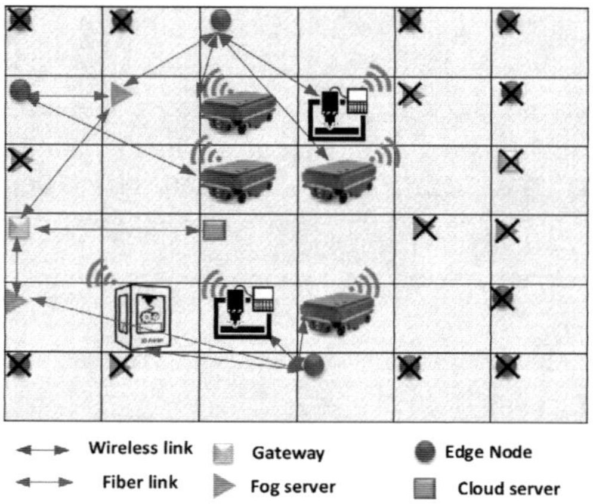

Figure 3. Optimized nodes positions.

5. Acknowledgments

Authors thank Cisco for funding this work conduced within the education and research chair between Cisco, CESI and VINCI Energies.

6. References

[1] Kagermann, H., Helbig, J., Hellinger, A., & Wahlster, W. (2013). Recommendations for implementing the strategic initiative INDUSTRIE 4.0: Securing the future of German manufacturing industry; final report of the Industrie 4.0 Working Group. Forschungsunion.

[2] Kang, H. S., Lee, J. Y., Choi, S., Kim, H., Park, J. H., Son, J. Y.,.. & Do Noh, S. (2016). Smart manufacturing: Past research, present findings, and future directions. International Journal of Precision Engineering and Manufacturing-Green Technology, 3(1), 111-128.

[3] Khan, W. A., Wisniewski, L., Lang, D., & Jasperneite, J. (2017, June). Analysis of the requirements for offering industrie 4.0 applications as a cloud service. In 2017 IEEE 26th International Symposium on Industrial Electronics (ISIE) (pp. 1181-1188). IEEE.

[4] Chang, C., Srirama, S. N., & Buyya, R. (2017). Indie fog: An efficient fog-computing infrastructure for the internet of things. *Computer*, 50(9), 92-98.

[5] O'Donovan, P., Gallagher, C., Leahy, K., & O'Sullivan, D. T. (2019). A comparison of fog and cloud computing cyber-physical interfaces for Industry 4.0 real-time embedded machine learning engineering applications. Computers in Industry, 110, 12-35.

[6] Bouzarkouna, I., Sahnoun, M. H., Sghaier, N., Baudry, D., & Gout, C. (2018, August). Challenges facing the industrial implementation of Fog Computing. In 2018 IEEE 6th International Conference on Future Internet of Things and Cloud (FiCloud) (pp. 341-348). IEEE.

[7] Bill, L. (2016). ORGIndustry 4.0: Intelligent and flexible productionDigitization.

[8] Bar-Magen Numhauser, J., & Guitierrez de Mesa, J. A. (2013). XMPP distributed topology as a potential solution for fog computing. In MESH 2013 the sixth international conference on advances in mesh networks.

[9] Qu, T., Lei, S. P., Wang, Z. Z., Nie, D. X., Chen, X., & Huang, G. Q. (2016). IoT-based real-time production logistics synchronization system under smart cloud manufacturing. The International Journal of Advanced Manufacturing Technology, 84(1-4), 147-164.

[10] Lin, C. C., & Yang, J. W. (2018). Cost-efficient deployment of fog computing systems at logistics centers in industry 4.0. IEEE Transactions on Industrial Informatics, 14(10), 4603-4611.

AUTHOR INDEX

Institute of Physics
Dirac House, Temple Back
Bristol BS1 6BE UK

ISSN: 1757-8981
ISBN 978-1-7138-2806-8

CURRAN ASSOCIATES INC.
proceedings.com